Third International Symposium on Pre-Harvest Sprouting in Cereals

Also of Interest

*Science, Agriculture, and the Politics of Research, Lawrence Busch and William B. Lacy

Proceedings of the XIV International Grassland Congress, edited by J. Allan Smith and Virgil W. Hays

Crop Reactions to Water and Temperature Stresses in Humid, Temperate Climates, edited by C. David Raper, Jr., and Paul J. Kramer

Azolla as a Green Manure: Use and Management in Crop Production, Thomas A. Lumpkin and Donald L. Plucknett

The Role of Centrosema, Desmodium, and Stylosanthes in Improving Tropical Pastures, edited by Robert L. Burt, Peter P. Rotar, and James L. Walker

Tomatoes in the Tropics, Ruben L. Villareal

Wheat in the Third World, Haldore Hanson, Norman E. Borlaug, and R. Glenn Anderson

Energy Analysis and Agriculture: An Application to U.S. Corn Production, Vaclav Smil, Paul Nachman, and Thomas V. Long

Agriculture as a Producer and Consumer of Energy, edited by William Lockeretz

Future Dimensions of World Food and Population, edited by Richard G. Woods

*Available in hardcover and paperback.

About the Book and Editors

Third International Symposium on
Pre-Harvest Sprouting in Cereals
edited by James E. Kruger and Donald E. LaBerge

Pre-harvest sprouting of cereals results in large economic losses each year for producers around the world. This volume, an outgrowth of the 1982 symposium held in Winnipeg, Canada, presents scientific research conducted internationally since 1978 and geared toward solving the problem of pre-harvest sprouting. Its multidisciplinary approach incorporates the perspectives of plant breeders, physiologists, and biochemists to provide a rounded discussion of the possible alternatives.

Dr. Kruger is a research scientist and head of the wheat enzyme section in the Grain Research Laboratory of the Canadian Grain Commission in Winnipeg, Manitoba. *Dr. LaBerge* is also a research scientist and is head of malting barley research in the Grain Research Laboratory.

International Organizing Committee

President

Dr. M. D. Gale
Plant Breeding Institute
Maris Lane, Trumpington
Cambridge CB2 2LQ
England

Secretary

Dr. J. E. Kruger
Grain Research
Laboratory
Canadian Grain
Commission
1404-303 Main Street
Winnipeg, Manitoba
R3C 3G8, Canada

Assistant Secreta.

Dr. J. Noll
Agriculture Canada
Research Station
195 Dafoe Road
Winnipeg, Manitoba
R3T 2M9, Canada

Members

Dr. V. Stoy
The Swedish Seed Association
S-268 00 Svalöv
Sweden

Dr. F. Weilenmann
Swiss Federal Research Station
    for Agronomy
Zurich-Reckenholz
Switzerland

Dr. N. F. Derera
The University of Sydney
Plant Breeding Institute
P.O. Box 219
Narrabri NSW 2390
Australia

Dr. L. Briggle
Science and Education Adm.
Agricultural Research, USDA
Beltsville, MD 20705
U.S.A.

Local Organizing Committee

Dr. J. E. Kruger
Dr. D. E. LaBerge
Dr. B. A. Marchylo
Grain Research Laboratory
Canadian Grain Commission
1404-303 Main Street
Winnipeg, Manitoba
R3C 3G8, Canada

Dr. J. Noll
Agriculture Canada
Research Station
195 Dafoe Road
Winnipeg, Manitoba
R3T 2M9, Canada

Third International Symposium on Pre-Harvest Sprouting in Cereals

edited by James E. Kruger
and Donald E. LaBerge

Routledge
Taylor & Francis Group

LONDON AND NEW YORK

First published 1983 by Westview Press

Published 2019 by Routledge
52 Vanderbilt Avenue, New York, NY 10017
2 Park Square, Milton Park, Abingdon, Oxon OX14 4RN

Routledge is an imprint of the Taylor & Francis Group, an informa business

Library of Congress Catalog Card Number 83-60546

ISBN 13: 978-0-367-27395-8 (hbk)
ISBN 13: 978-0-367-27433-7 (pbk)

Contents

Preface

This volume represents the proceedings of the Third International Symposium on Pre-Harvest Sprouting Damage in Cereals, held
at Hecla Island, Manitoba, Canada, in June 1982. The purpose of
the meeting was to provide a forum for plant breeders, biochemists,
physiologists, and technologists to review the current status of
pre-harvest sprouting and to exchange new information and ideas
related to solving this problem, which is responsible for large
economic losses in cereals each year.

The background for research in this area goes back to 1973,
when it was recognized that there was not enough interaction among
research workers studying pre-harvest sprouting. An organizing
committee was formed in that year, with N. Derera as chairman, to
help facilitate the interchange of ideas and to assist in the coor-
dination of research of mutual interest. The result was a highly
successful symposium in 1975 in Rostanga, Sweden, organized by Dr.
V. Stoy of the Swedish Seed Association, with eleven nations
participating. The proceedings were published in Cereal Research
Communications, Volume 4(2), 1976. The symposium was a catalyst for
greatly expanded research. Thus the pre-harvest sprouting problems
were clearly defined, researchers became more aware of each other's
work, and mutually beneficial interdisciplinary projects were
planned. To accommodate new research and ideas, a second symposium
was hosted by Dr. M. Gale of the Plant Breeding Institute, Cambridge,
England, in 1979, with sixteen nations participating. Although
larger than the previous symposium, the meeting maintained the workshop format and demonstrated some notable breakthroughs on the pre-
harvest sprouting damage problem. Proceedings of the meeting were
again published in Cereal Research Communications, Volume 8(1),
1980.

The current meeting maintained the same successful workshop
format, with the program being broken into three principal sections,
all represented in this volume. Following an introductory discourse
on the progress made in pre-harvest sprouting research, by Dr. V.
Stoy of the Swedish Seed Association, the sections deal with the
physiology, biochemistry, and plant breeding aspects of the sprouting problem. Of course, overlapping between the disciplines can be
found in many of the papers. Although research in the past has
concentrated mainly on the cereals, wheat and barley, this

symposium also had a sub-session devoted to the increasing concern regarding the pre-harvest sprouting problem in maize.

At the closing banquet, recognition in the form of a plaque was given to Mr. N. Derera, plant breeder, for his outstanding contribution over the years in making progress toward solving the pre-harvest sprouting problem.

The organizing committee is grateful to many people for making the conference a success. There are the many local volunteers to be thanked, including E. Czarnicki, N. Turriff, D. Levich, and L. Morris. A special thanks is extended to J. Ramsay, D. Chance, and H. Zimberg for assistance in readying these proceedings for publication.

Mr. H.D. Pound, Chief Commissioner of the Canadian Grain Commission, and Dr. K.H. Tipples, Director of the Grain Research Laboratory were kind in allowing extensive use of the manpower and facilities of the Canadian Grain Commission. All participants, whether speakers, session chairmen, or part of the audience, provided the necessary enthusiasm.

The symposium was sponsored financially and in other ways by a number of organizations. The organizing committee wishes to acknowledge the generous support of:

> Australian Wheat Board, Australia
> Canadian Grain Commission, Canada
> Canadian International Grains Institute, Canada
> Canadian Wheat Board, Canada
> Doty Laboratories, U.S.A.
> New South Wales Flour Millers' Council, Australia
> Norwegian Grain Corporation, Norway
> Prime Wheat Association Limited, Australia
> Swedish Grain Commission, Sweden

James E. Kruger
Donald E. LaBerge

Introduction

Progress and Prospect in Sprouting Research

Volkmar Stoy, *The Swedish Seed Association, Svalöv, Sweden*

INTRODUCTION

At the very beginning of this 3rd International Sprouting Symposium it is a natural reaction to look back at the preceding 2nd Symposium at Cambridge, England, three years ago and to make a short summing up of the major results and impacts of that conference. In this way a useful starting line is set for the forthcoming lectures and discussions. The introductory talk at the Cambridge meeting was given by N.F. Derera and his presentation and description of the sprouting problem is in essence still valid why I can add only little to his message. I would like, however, to stress once again the great economic losses sprouting damage may cause some years in various regions of the world. The global importance of the sprouting problem is perhaps not always realized but, in fact, all continents and all cereal crops may be effected more or less frequently. It is significant for the increasing awareness of the problem, however, that a whole session of the 3rd symposium is devoted to maize, a crop which hitherto has not been seriously considered in this context.

It is not easy to obtain reliable figures of the total damage that may be caused during one year to the world's grain crops but there are many examples from various regions that certain crops can be very seriously hit in special years and that 30-50 per cent or even more of the grains harvested during such a year may be so severely damaged that they are unsuited for human consumption.

Another point that deserves to be mentioned is that sprouting damage often occurs erratically and non-predictably. This is not surprising since readiness to sprouting is the resultant of the genetic constitution of the grain and the influence of the weather both before and during the harvest season. As the weather may vary considerably from year to year - often in a cyclic fashion - in many regions of the world, years with good harvest conditions are not seldom followed by periods of bad weather and vice versa. As a consequence of this a

general interest in the problem flares up for a few years
under the influence of a bad weather period only to die
off again as soon as the climatic situation improves.
Research work suffers definitely from such a fluctuating
interest and it must be strongly emphasized that a condition for a successful work on this complicated problem is
that research can be planned and carried through on a
reasonable scale also in the intervening "favourable"
period.

At the Cambridge-symposium in 1979 the lectures and
discussions were focused on three major themes: the
chemistry, the physiology and the breeding and genetic
aspects of pre-harvest sprouting. A similar approach has
been adopted at the present meeting and it may therefore
be appropriate to start with a short review of the status
within each section as we left it in 1979.

CHEMISTRY

Looking back at the reports from different laboratories on work made on the chemistry aspect of the sprouting
problem it is quite clear that substantial progress has
been made in this field during the late 70ies. Particularly the properties of different types of α-amylase isoenzymes have been investigated in great detail at several
laboratories using different advanced biochemical
techniques. Thus the early electrophoretic investigations
of Olered and Jönsson have been refined and deepened
considerably in the late seventies as demonstrated by the
reports of Kruger and collaborators. Their results were
supported in many respects by the very elegant immunochemical studies of Daussant and co-workers and of
Sargeant. It is now clear that there exist two major
groups of α-amylase isoenzymes in wheat and that these
two groups of isoenzymes appear to be analogous to
corresponding groups in other cereals such as barley and
triticale. The abundance of these groups varies with the
stage of grain development and their synthesis seems to
be controlled by some factors in the embryo, but as late
as in 1979 the mechanism of this control was not clear.

The characterization of the numerous isoenzymes,
revealed by modern analytical techniques, is still in
many respects an unsolved problem. Moreover, it is uncertain how results obtained in in vitro experiments
reflect the reactions in vivo. I also would like to draw
attention to the possible existence of enzyme inhibitors,
in the grains, a very little investigated but inspiring
problem for biochemists and physiologists.

Another question, which emerges out of these consi-
derations, concerns the physiological role of these
different amylase isoenzymes. Do they only function as
germination enzymes or do at least some of them also play
an important role during grain filling and if so, what is
the exact mechanism? Hopefully more information on these

and related problems will be presented during the lectures
and discussions of the next days.

Equally important as a thorough knowledge of the
types and properties of different hydrolytic enzymes
active in the germination process is reliable information
about the sites of their synthesis and activation and
about their relase and transport into the endosperm part
of the grains. One of the most spectacular lectures at the
previous symposium was that of <u>Gregory Gibbons</u> who presented strong evidence that at least in barley α-amylase is
synthesized not only in the aleurone cells, as generally
believed, but also in the scutellum tissues. Gibbons and
his groups in Copenhagen used an elegant immunofluore-
scence technique and could show that during the first
days of germination α-amylases moved out as a front from
the scutellum and continued into the endosperm and that,
in fact, α-amylase was not produced in the aleurone cells
in significant quantities until the third day after
germination started. The exciting work has been continued
since then and it is obvious that the traditional view of
amylase-production in the germinating grain has to be
revised considerably.

PHYSIOLOGY

The physiology of the control of dormancy and germination is a field of obvious importance in this context.
<u>Ian Gordon</u> presented a very stimulating introduction to
this point at the Cambridge-conference and gave his very
personal views on this obviously rather complicated
subject. He rightly emphasized the importance of the early
metabolic events which regulate the germinability of the
grain and concluded that the physiology of sprouting
damage is mainly equivalent to the physiology of germination and dormancy which includes problems such as enzyme
activity, hormonal interactions, the provision of energy
and substrates, etc. Particular emphasis was laid on the
hypothesis that initial hypo-oxia exists in the grain and
that a reversion of this hypo-oxia by some mechanism or
other is one of the crucial first steps in the onset of
germination.

It became quite clear both from this introduction
and from the subsequent more specialized lectures and
discussions that the physiology of dormancy and germination still is "terra incognita" in many respects and that
much more basic research work is needed until a reasonably
true picture of the metabolic events and the control
mechanisms is achieved. This lack of knowledge also
explains why it is so difficult to work out good and
simple selection methods based on "early" well-defined
biochemical or physiological characters, as substitutes
to the hitherto commonly used "end-result"-methods.

BREEDING AND GENETIC ASPECTS

Selection based on dormancy criteria of non-stressed grains is one step in this direction and methods using this approach were reported already at the first symposium by Olsson and Mattsson and now again by Strand. With these methods the potential resistance of the material can be estimated but for a final test of cultivars or advanced breeding material a combination of exposure to severe sprouting conditions and subsequent evaluation of the damage by determination of α-amylase activity or falling number still seems to be the most widely used procedure as reported by Weilenmann, Strand, and Plarre.

Another practical problem concerns the finding of suitable genes for sprouting resistance and their intro- duction into current breeding material. It was quite clear from the contributions devoted to this problem that there exists a substantial genetic variability in most cereals with respect to susceptibility to sprouting and also that different kinds of gene effects are involved. Thus for instance in wheat many of the identified genes for sprouting resistance are associated with red colour of the seed coat (Reitan, McEwan) but there also exist other genes which can induce considerable dormancy even in white-seeded cultivars (Bhatt and Derera). A thorough screening of existing collections of wild types and cultivated forms will certainly reveal further sources of useful genes for the improvement of sprouting resistance along different pathways.

OUTLOOK INTO THE FUTURE

What, then are the prospects for future research work in the field of sprouting resistance? I think it is justified to be rather optimistic in this respect, provid- ed that continued personal and financial resources are available. The lectures and discussions during this week will certainly show that substantial advances have been made since 1979 and it is my personal belief that we will witness further rapid progress in the field of germination chemistry and biochemistry parallel with a deepened in- sight into the physiological mechanisms which regulate the basic metabolic events in the sprouting process. Thus, our increased knowledge in the field of phytohormone phys- iology will undoubtedly contribute very much to a better understanding of the control of the critical first steps in germination.

As a consequence of these achievements it should be possible to develop new systems in the breeding for resis- tance to sprouting damage, to disclose new gene sources to be used in this work and to specify the inheritance of these genes in considerable detail. Hopefully new analytical tools will be developed at the same time which enable the breeders to screen large materials for the desired genes rapidly and at reasonable costs. This latter aspect is of course also of great interest to the different grain consumers. They, too, need cheap and rapid methods to evaluate the quality of the products offered to them. It is our task to help the industry to develop these tools by specifying what to look for and I am sure the challenge will be met efficiently.

Section I
Physiology of Pre-Harvest Sprouting

The Physiology of Pre-Harvest Sprouting—A Review

R. W. King, CSIRO, Division of Plant Industry, Canberra, Australia

INTRODUCTION

It is probable that selection was against sprouting during the domestication of cereals. However, sprouting still remains a problem and this is testimony to how little we know about the processes which regulate germination. We can only echo now what Theophrastus said over two thousand years ago:

"the causes of such differences (in germination) must be found in several different circumstances" after Evenari (1980/81).

Efforts to produce more marketable seed, fruit or other products repeatedly highlight the inadequacies of physiological understanding of germination. In celery, for example, breeding against bolting led to greater seed dormancy (Thomas, 1978). In tomato, sprouting inadvertently became a problem in a line bred for high yield for once-over harvesting of canning fruit (Santos and Yamaguchi, 1979). In cereals, selection for high-protein and vigorous seed may have favoured sprouting as must the rapid turn around between harvest and resowing of seed which is common practice today.

In the context of this symposium the role for physiology is, therefore, three-fold; to provide an understanding of the regulation of germination; to develop selection criteria for the plant breeder and; to predict changes in sprouting which will result from various breeding strategies. Since the last symposium there has been considerable progress in all these areas and this recent information and some of the older literature is reviewed below.

THE PHYSIOLOGY OF DORMANCY IN CEREALS

Dormancy induction

In their review Taylorson and Hendricks (1977) stated that dormancy in seeds "is an arrest of development of seed embryos". Abscisic acid (ABA) may be involved here (see Duffus this symposium and King, 1982).

For instance, applied ABA blocks germination whilst
maintaining embryo development (Umbeck and Norstog, 1979).
Endogenous levels of ABA are also elevated over the
period that embryos first become dormant (King, 1976) and
at the time that specific storage proteins are synthesized
(Sussex and Dale, 1979). As a corollary, low ABA levels
are associated with viviparous germination in a number of
the single gene mutants of corn (Brenner et al., 1977;
Smith et al., 1978; Robichaud, 1979). Vivipary also
results in normal corn treated with Fluridone - a
carotenoid and, possibly ABA synthesis inhibitor (Fong,
this volume). Similarly, promotion of germination of
immature grain by drying (Evans et al., 1975; King,
1976; Nicholls, 1979) is associated with enhanced
degradation of endogenous ABA (King, 1976; 1979) and of
applied ^{14}C-labelled ABA (King, 1979). Clearly, dormancy
induction may occur well before harvest and ABA may be
implicated in this control although there are many other
explanations (see Nicholls, 1979).

Maintenance of dormancy
         Three aspects of dormancy maintenance are considered
below: physical barriers, chemical controls and metabolic
regulation.
         Physical barriers may be responsible for the
maintenance of dormancy. Wellington (1956), for example,
proposed that the dormant, red-grained wheats have a more
restrictive seed coat than the white-grained, non-dormant
wheats. He found that initial (24h) water absorption by
the'embryo was similar for dormant and non-dormant grain.
However, further water uptake by the embryo was delayed
in the dormant variety unless the embryo was uncovered or
the distal half of the seed was removed. This he argued
indicated differences in seed coat constraint (not of
permeability) between red and white-grained varieties.
However, other explanations cannot be excluded. The
embryos themselves may have differed in ability to grow.
Also, the seed coat could have blocked oxygen uptake by
the embryo (see Gordon, 1980) although others have discarded this possibility (e.g. Chao et al., 1959). A
further suggestion is that the seed coats are a source of
germination inhibitor(s) (see Miyamoto and Everson, 1959)
and/or a barrier to the leaching of inhibitor(s).
Clearly, to answer these suggestions further examination
is required of the differences in dormancy of red- and
white-grained wheats.
         Chemical control over dormancy may involve plant
growth regulators. Cereal grain germination is influenced by treatment with the known growth regulators (Khan,
1980/81) and so they may act naturally either in dormancy
maintenance (inhibitors) or in dormancy termination.
However, this viewpoint has yet to be substantiated.
Measurements of endogenous levels of ABA by Berrie and

co-workers (1979) showed no relationship between ABA content and dormancy in wild oat. The same appears to be true for wheat grain (see King, 1976; 1982) and, since most of the ABA present in immature grain is degraded at maturity (e.g. King, 1976), it is unlikely that the lower content of ABA in mature grain would be sufficient to impose dormancy unless it is all concentrated in the embryo. However, Goldbach and Michael (1976) detected small ABA differences in mature grain of dormant and non-dormant barley varieties.

For a chemical controlling dormancy, then, with release of dormancy, its content might be expected to decline if it is an inhibitor or increase if it is a promoter. Studies of Berrie et al. (1979) with wild oat (Avena fatua) illustrate possible inhibitor/decay regulation of dormancy by C_7, C_8 and C_9 fatty acids. These compounds are more abundant in dormant than in non-dormant seed. They inhibit germination and the most potent, the C_9 compound decreases in the air dry seed either by evaporation or degradation with a half life of 56.5 days. This study provides strong circumstantial evidence that fatty acids could act in an inhibitor/decay mode of dormancy control.

Dormancy based on the absence of a promoter could result from a restriction in the synthesis of a compound such as gibberellin, as appears to be the case in dormant hazel seeds (e.g. Williams et al., 1974). Alternatively, there could be changes in degradation of this promotive hormone. For example, in developing wheat grain, gibberellin synthesis is maintained but, in contrast to vegetative tissue, novel and probably inactive gibberellins are formed (Gaskin et al., 1980). In barley, exogenously supplied GA_3 is rapidly glycosylated and so inactivated (Smith and Briggs, 1980). There is as yet no evidence to link these controls over GA level with dormancy but it is clear that we need to consider such controls and not only how they are imposed but also how they are terminated.

At the level of metabolic control of dormancy, germination may be associated with more rapid catabolism of glucose-6-phosphate via the pentose phosphate pathway than by glycolysis (see Taylorson and Hendricks, 1977; Gordon, 1980; Duffus, this volume) and GA may trigger these changes (Simmonds and Simpson, 1971). However, recent studies by Adkins and Ross (1981) with Avena fatua germinated at 25° or imbibed at $5^{\circ}C$ have shown that there is no essential connection between loss of dormancy and increase in the activity of the pentose phosphate pathway. Activation of this pathway is probably a consequence of rather than a cause of loss of dormancy.

SPROUTING AND ENVIRONMENT
Temperature, light and nutrition

Many aspects of the environment and of nutrition during grain development influence subsequent germinability and ability of grain to synthesize α-amylase and other hydrolytic enzymes. As shown in Table 1 control mechanisms are only known in some instances.

TABLE 1. Effect of environmental and trophic factors during seed development on germination processes and sprouting.

Environmental Factors	Seed Response	Author
Humidity Temperature Photoperiod	Altered dormancy and sprouting	Grahl and Schrödter(1975) Gutterman (1980/81) Takahashi (1980)
Temperature	GA-induced amylase synthesis	Nicholls (1980)
Temperature Light quality	Switch to GA-independent amylase synthesis	Nicholls (this volume)
Nutritional Factors		
Low molybdenum (enhance with NO_3)	Sprouting in maize	Tanner (1978)
High NO_3	More diastatic activity	Hayter and Riggs (1972)
High NO_3	More amylase on germination	Ching and Rynd (1978) Huang and Marston (1980)

The effect of light and of photoperiod on germination of seeds of some species may involve the pigment phytochrome. If phytochrome is not present in the active P_{fr} form on seed drying then germination will be light dependent. For cereal seed there appears to be no light requirement for germination (see Toole, 1973). However, in seeds of some species high levels of chlorophyll and other pigments in the seed coat act as a screen for incident light and so impose a light requirement for germination (Cresswell and Grime, 1981). Thus, a light requirement for germination of cereals might be introduced by selecting for pigmented seed coats.

Effects of environment on sprouting probably involve complex changes. However, it may be possible to

identify controlling processes. For example, high
temperatures (Radley, 1976) and reduced availability of
potassium (Haeder and Beringer, 1981)   influence   ABA
contents in wheat grain. Water stress might also result
in a similar increase in grain ABA (Goldbach and Goldbach,
1977). It has yet to be shown whether these changes in
ABA induced by environment and nutrition influence
germinability but further examination is clearly
warranted.

A more compelling illustration of processes controlling sprouting is the relationships between
nutrition and sprouting demonstrated by Tanner (1978).
By utilizing acid soils (pH 4.3-4.7) he showed that the
molybdenum content of the corn kernel was critical for
sprouting. Sprouting occurred at levels below 0.05 ppm
and this could be blocked by applications of molybdenum.
The explanation proposed was that nitrate, a germination
promoter, had built up in the low molybdenum kernels and,
as predicted, application of nitrogenous fertilizer
exacerbated the sprouting problem.

An explanation is not as obvious for the effects of
high levels of nitrogen fertilizer on amylase production
during germination of wheat and on development of
diastatic activity during malting of barley (Table 1).
Hayter and Riggs (1972) suggested that the increased
diastatic power of barley reflected an increase in β-
and not α-amylase. As β-amylase synergizes α-amylase
action (Briggs, 1962) the same explanation may hold for
the studies with wheat. An alternative is that aleurone
size or function was stimulated directly in the high-
protein wheat grains. Whatever the answer, it is clear
that enhanced amylase activity might be a consequence of
breeding wheat of higher grain protein content and
breeders should be aware of this conflict of objectives.

Rainfall

To control sprouting by restricting grain water
uptake is simple when seed coat impermeability (hard
seededness) can be selected as in legumes such as soybean
(Kilen and Hartwig, 1978). However, in cereals seed coat
impermeability is not known. The study of Wellington
(1956), cited above, indicated that the differences he
found in embryo water uptake after 48h of wetting were
not related to seed coat permeability but were caused by
differences in germination. Nevertheless, varietal
differences in grain wetting can be detected (Butcher and
Stenvert, 1973; Clarke, 1980; King, in preparation)
when care is taken to distinguish physical processes
during early grain imbibition (e.g. up to 24h of wetting)
from subsequent germination. Butcher and Stenvert (1973)
found slowest water penetration into grain of the high
protein, hard, white, spring wheat Timgalen. This
cultivar also had a thick outer cuticle and testa (Moss,

1973). Conversely, most rapid water penetration was
found for Heron, a soft-grained cultivar with low protein
and thin testa. It is not known if these grain characteristics correlate with water penetration rates in a larger
grouping of cultivars or if differences in grain water
uptake correlate with sprouting. However, when whole
ears are wet, differences in the rate of wetting do
influence sprouting (King and Chadim, this volume).

Obviously advances can be made by identifying simple
characters controlling ear and grain wetting. Control of
sprouting may be one step closer if all such characters
were incorporated into one variety. Even so, because of
the variability of natural rainfall and in the subsequent
evaporation of water, reduced rates of grain wetting may
not be a complete answer in some environments. For
example, slow ear wetting may not be advantageous under
brief intermittent showers because both wetting and drying
could be slowed. Nevertheless, studies of wetting and
drying cycles with wheat grain have established that
rewetting is more rapid as is germination (Lush et al.,
1981). Moreover, provided that embryo development has
not advanced to the point of no return - visible sprouting - each hydration-dehydration event can be summated.
Within this limitation the grain recommences from where
it left off in the previous rain shower (Akalehiywot and
Bewley, 1980; Lush et al., 1981). Thus, under most
field conditions slower ear and grain wetting could be
beneficial in the control of sprouting.

THE MARGINS OF PHYSIOLOGY

Biochemistry

Regulation of α-amylase production is one important
interface between biochemical and physiological studies.
In wheat, only a small amount (< 1%) of the α-amylase
produced during germination may be the "green amylase" of
the immature seed (Olered and Jonsson, 1970). Its
activity may also be reversibly regulated by wetting and
drying in the field (Olered and Jonsson, 1970). During
germination the bulk of the α-amylase is newly synthesized whilst for β-amylase there may be both synthesis
(Okamoto and Akazawa, 1980) and activation-release.

We now have extensive knowledge of molecular and
hormonal aspects of the regulation of synthesis of α-
amylase. Both GA and ABA are implicated (see Duffus,
this volume). As a consequence, it has been possible to
investigate hormonal mechanisms of regulation. For
instance, by using in vitro mRNA translation techniques
it has been shown that massive synthesis of α-amylase at
maturity in the grain of triticale 6A190 is associated
with production of α-amylase mRNA at the time (King et al
1979). Also, triggering of this α-amylase synthesis is
possible with GA_3 as is known for germination of normal,
non-sprouted grain. Thus, this defect in triticale 6A190

apparently involves premature GA production by the embryo just before amylase synthesis and sprouting occur. Unlike viviparous sprouting in corn which may correlate with lowered ABA levels (see above) in triticale 6A190 ABA content of grain was no lower than in normal wheat (King et al., 1979).

In vitro translation techniques do not offer the sensitivity of assay required to determine when new amylase mRNA is first formed in non-sprouted grain. However, recent advances in recombinant DNA technology may allow more sensitivity in assaying mRNA and could provide answers to questions of the nature and timing of synthesis of "green amylase".

Genetics

Our understanding of the physiology of sprouting can be rapidly advanced by the use of genetic variation for identifying component processes of sprouting. For example Ho and co-workers (1980) have begun screening for amylase production mutants of barley which are blocked for gibberellin and ABA response. It can be anticipated that these mutants will not all be defective in the same biochemical step. Therefore, it should be possible to build up a library of mutants for the various steps controlling the final response, α-amylase production.

A complementary approach is to select for lines with altered hormone level. Rice lines are known which under-produce GA and ABA (Suzuki et al., 1981). Some dwarf corn lines also under-produce GA although other dwarf lines of corn and wheat (Gale, 1976) are insensitive to GA. Wheat lines producing low or high levels of ABA in the leaf have also been reported recently by Quarrie (1981). These lines offer excellent material for a study of the roles played by ABA and GA in sprouting and α-amylase production.

A recent illustration of the value of genetics in unravelling physiological controls is seen in the study of King and Chadim (this volume) on the effects of awns on sprouting in wheat. It was the availability of near-isogenic awned and awnless lines that allowed the conclusion that varietal differences in ear wetting were related to the possession of awns.

The relationship between seed coat colour and germinability is one area where physiological and genetic studies could be further extended. Stoy and Olsen (1980) reported simple genetic control of sensitivity of embryos of wheat to catechin-tannin extracts. Also, coat colour is simply inherited (see Gordon, 1980). However, physiologists have yet to examine, further, the report of Miyamoto and Everson (1958) that red-grained wheat contains more catechin-tannins. We still lack compelling evidence that red coat colour is involved with chemical inhibition of sprouting.

CONCLUSION

Despite the fascination and excitement of studies of grain sprouting, our understanding of dormancy is quite incomplete. Only when we know more about the physiological and biochemical processes controlling germination will it be possible to make intelligent attempts to manipulate grain sprouting. Meanwhile, the lack of this physiological understanding could result in quite unexpected problems with sprouting if breeders select for such simple characteristics as seed size, protein content and milling characteristics.

REFERENCES

Adkins, S.W. and Ross, J.D., 1981. Studies in wild oat seed dormancy. II. Activities of pentose phosphate pathway dehydrogenases. Plant Physiol. 68, 15-17.

Akalehiywot, T. and Bewley, J.D., 1980. Desiccation of oat grains during and following germination and its effects upon protein synthesis. Can. J. Bot. 58, 2349-2355.

Berrie, A.M.M., Buller, D.R., Don, R. and Parker, W., 1979. Possible role of volatile fatty acids and abscisic acid in the dormancy of oats. Plant Physiol. 63, 758-764.

Briggs, D.E., 1962. Gel-diffusion method for the assay of α-amylase. J. Inst. Brewing 68, 27-32.

Butcher, J. and Stenvert, N.L., 1973. Conditioning studies on Australian wheat III. The role of the rate of water penetration into the wheat grain. J. Sci. Fd. Agric. 24, 1077-1084.

Chao, T.F., Wang, F.T., Wang, S., and Tsai, T.P., 1959. A study on the physiology of after ripening of wheat grains. Acta Botanica Sinica 8, 230-238.

Ching, T.M. and Rynd, L., 1978. Developmental differences in embryos of high and low protein wheat seeds during germination. Plant Physiol. 62, 866-870.

Clarke, J.M., 1980. Measurement of relative water uptake rates of wheat seeds using agar media. Can. J. Plant Science 60, 1035-1038.

Cresswell, E.G. and Grime, J.P., 1981. Induction of a light requirement during seed development and its ecological consequences. Nature 291, 583-585.

Evenari, M., 1980/81. The history of germination research and the lesson it contains for today. Israel J. Botany 29, 4-21.

Evans, M., Black, M., and Chapman, J., 1975. Induction of hormone sensitivity by dehydration is one positive role for drying in cereal seeds. Nature 258, 144-145.

Gale, M.D., 1976. High α-amylase-breeding and genetical aspects of the problem. Cereal Res. Comm. 4, 231-243.

Gaskin, P., Kirkwood, P.S., Lenton, J.R., MacMillan, J. and Radley, M.E., 1980. Identification of gibberellins in developing wheat Triticum aestivum grain. Agric. Biol. Chem. 44 (7), 1589-1594.

Goldbach, H. and Goldbach, E., 1977. Abscisic acid translocation and influence of water stress on grain abscisic acid content. J. Exptl. Bot. 28, 1342-1350.

Goldbach, H. and Michael, G., 1976. Abscisic acid content of barley grains during ripening as affected by temperature and variety. Crop Science 16, 797-799.

Gordon, I.L., 1980. Germinability, dormancy and grain development. Cereal Res. Comm. 8, 115-129.

Grahl, A. and Schrödter, H., 1975. Witterung vor der Reife und Keimruhe von Weizen unter besonderer Berucksichtigung der Auswuchsvorhersage. Seed Sci. and Technol. 3, 815-826.

Gutterman, Y., 1980/81. Influences on seed germinability : phenotypic maternal effects during seed maturation. Israel J. Bot. 29, 105-117.

Haeder, H.E. and Beringer, H., 1981. Influence of potassium nutrition and water stress on the content of abscisic acid in grains and flag leaves of wheat during grain development. J. Sci. Fd. Agric. 32, 552-556.

Hayter, A.M. and Riggs, T.J., 1972. Environmental and varietal differences in diastase power and four associated characteristics of spring barley. J. Agric. Sci. 80, 297-302.

Ho, T.H.D., Shih, S-C., and Kleinhofs, A., 1980. Screening for barley mutants with altered hormone sensitivity in their aleurone layers. Plant Physiol. 66, 153-158.

Huang, G. and Marston, E.V., 1980. α-amylase activity and preharvest sprouting damage in Kansas hard white wheat. J. Agric. Fd. Chem. 28, 509-512.

Khan, A.A., 1980/81. Hormonal regulation of primary and secondary seed dormancy. Israel J. Bot. 29, 207-224.

Kilen, T.C. and Hartwig, E.E., 1978. An inheritance study of impermeable seed in soybeans. Field Crops Res. 1, 65-70.

King, R.W., 1976. Abscisic acid in developing wheat grains and its relationship to grain growth and maturation. Planta 132, 43-51.

King, R.W., 1979. Abscisic acid synthesis and metabolism in wheat ears. Aust. J. Plant Physiol. 6, 99-108.

King, R.W., 1982. Abscisic acid in seed development. Chapter 7 in A.A. Khan (ed.) The physiology and biochemistry of seed development, dormancy and germination. Elsevier/North-Holland Biomedical Press. (In press)

King, R.W., Salminen, S.O., Hill, R.D. and Higgins,
T.J.V., 1979. Abscisic-acid and gibberellin action
in developing kernels of triticale (cv. 6A190).
Planta 146, 249-55.

Lush, M.W., Groves, R.H., and Kay, P.E., 1981. Presowing
hydration-dehydration treatments in relation to seed
germination in early seedling growth of wheat and
ryegrass. Aust. J. Plant Physiol. 8, 409-425.

Miyamoto, T. and Everson, E.H., 1958. Biochemical and
physiological studies of wheat seed pigmentation.
Agron. J. 50, 733-734.

Moss, R., 1973. Conditioning studies on Australian
wheat II. Morphology of wheat and its relationship
to conditioning. J. Sci. Fd. Agric. 24, 1067-1076.

Nicholls, P.B., 1979. Induction of sensitivity to
gibberellic acid in developing wheat caryopses :
effect of rate of desiccation. Aust. J. Plant
Physiol. 6, 229-240.

Nicholls, P.B., 1980. Development of responsiveness to
gibberellic acid in aleurone layer of immature
wheat and barley caryopses: effect of temperature.
Aust. J. Plant Physiol. 7, 645-653.

Okamoto, K. and Akazawa, T., 1980. Enzyme mechanism of
starch breakdown in germinating rice seeds 9. De
novo synthesis of β-amylase. Plant Physiol.
65, 81-84.

Olered, R. and Jonsson, G., 1970. Electrophoretic
studies of α-amylase in wheat II. J. Sci. Fd.
Agric. 21, 385-392.

Pool, M. and Patterson, F.L., 1958. Moisture relations
in soft red winter wheats II. Awned versus awnless
and waxy versus nonwaxy glumes. Agron. J. 50,
158-160.

Quarrie, S.A., 1981. Genetic variability and heritability of drought-induced abscisic acid
accumulation in spring wheat. Plant Cell and
Environ. 4, 147-151.

Radley, M., 1976. The development of wheat grain in
relation to endogenous growth substances. J.
Exptl. Bot. 27, 1009-1021.

Robichaud, C.S., 1979. An analysis of abscisic acid
relations in embryos of a viviparous mutant of Zea
mays. Plant Physiol. (Suppl.) 63 (5) p.36.

dos Santos, D. and Yamaguchi, M., 1979. Seed sprouting
in tomato fruits. Sci. Hortic. 11, 131-40.

Simmonds, J.A. and Simpson, G.M., 1971. Increased
participation of the pentose phosphate pathway in
response to after-ripening and GA treatment in
caryopses of Avena fatua L. Can. J. Bot. 49,
1833-1840.

Smith, J.D., McDaniel, S. and Lively, S., 1978.
Regulation of embryo growth by abscisic acid in
vitro. Maize Genetics Newsletter 52, 107-108.

Smith, M.T. and Briggs, D.E., 1980. Externally applied gibberellic acid and α-amylase formation in grains of barley (Hordeum distichon). Phytochem. 19, 1025-1033.

Stoy, V. and Olsen, O-A., 1980. Inheritance of a factor affecting the response to germination inhibitors in excised wheat embryos. Cereal Res. Comm. 8, 203-208.

Sussex, I.M. and Dale, R.M.K., 1979. Hormonal control of storage protein synthesis in Phaseolus vulgaris pp.129-141. In I. Rubenstein (ed.) The plant seed: development preservation and germination. New York, Academic.

Suzuki, Y., Kurogochi, S., Murofushi, N., Ota, Y. and Takahashi, N., 1981. Seasonal changes of GA_1, GA_{19} and abscisic acid in three rice cultivars. Plant and Cell Physiol. 22, 1085-1093.

Takahashi, N., 1980. Effect of environmental factors during seed formation on pre-harvest sprouting. Cereal Res. Comm. 8, 175-183.

Tanner, P.D., 1978. A relationship between premature sprouting on the cob and molybdenum and nitrogen status of maize grain. Plant and Soil 49, 427-432.

Taylorson, R.B. and Hendricks, S.B., 1977. Dormancy in seeds. Ann. Rev. Pl. Physiol. 28, 331-354.

Thomas, T.H., 1978. Relationship between bolting resistance and seed dormancy of different celery cultivars. Scientia Hortic. 9, 311-316.

Toole, V.K., 1973. Effects of light, temperature and their interactions on the germination of seeds. Seed Sci. and Technol. 1, 339-396.

Umbeck, P.F. and Norstog, K., 1979. Effects of abscisic acid and ammonium ion on morphogenesis of cultured barley embryos. Bull. Torrey Bot. Club 106, 110-116

Wellington, P.S., 1956. Studies on the germination of cereals 2. Factors determining the germination behaviour of wheat grains during maturation. Annals Bot. 20, 481-500.

Williams, P.M., Bradbeer, J.W., Gaskin, P. and MacMillan, J., 1974. Studies in seed dormancy VIII. The identification and determination of gibberellins A_1 and A_9 in seeds of Corylus avellana L. Planta 117, 101-108.

Gibberellic Acid Insensitivity Genes and Pre-Harvest Sprouting Damage Resistance

D. J. Mares, F. W. Ellison and N. F. Derera,
*The University of Sydney, Plant Breeding
Institute, Narrabri, New South Wales, Australia*

SUMMARY

A wide range of dwarf and semi-dwarf wheats were
grown at Narrabri in the north west of New South Wales
and their response to pre-harvest rain compared. The
response varied widely with both genotype and environment. Norin 10 dwarfing genes did not appear to
significantly influence the reaction to rain. Lines
containing *Gai*3/*Rht*3 were tolerant under most conditions,
but apparently susceptible under others. Whole grains
of a sample of Tordo (*Gai*3/*Rht*3), which appeared to be
sprouting susceptible, produced higher levels of
α-amylase during germination than a sprouting resistant
sample of the same variety. The ratio of the amount of
enzyme produced in the presence of added gibberellic acid
to that produced during germination in water was the same
in both cases. The implications of these results are
discussed.

INTRODUCTION

The Australian prime-hard wheat, Shortim, which
contains both the Norin 10 dwarfing genes (*Gai*1/*Rht*1 +
*Gai*2/*Rht*2) has shown moderate levels of tolerance to pre-
harvest sprouting damage over several years. This
observation stimulated an investigation of a wide range
of dwarf and semi-dwarf wheat varieties to see if
tolerance to weathering was associated with particular
dwarfing genes or gene combinations.

*Gai*3/*Rht*3, the dwarfing gene in some extreme dwarfs,
has already been connected with sprouting tolerance
(Gale and Marshall, 1975; Bhatt, Derera and McMaster, 1977
and Flintham and Gale, 1980). Unlike Norin 10 dwarfs,
where the insensitive response to gibberellic acid
appears to be confined to the aerial parts of the plant,
in *Gai*3/*Rht*3 dwarfs the grain aleurone is also un-
responsive to gibberellic acid. Gale and Marshall

(1975) reported that this gene was associated with very low levels of α-amylase in mature grain. When grown in the hot, dry environment of northern N.S.W., wheats containing *Gai3/Rht3* are unfortunately too short for commercial cultivation. However, in view of the paucity of sources of resistance in white-grained wheats, considerable effort has been expended in an effort to transfer the gibberellic acid insensitivity reaction to a plant of more acceptable height. To date this effort has been unrewarding.

The aim of this investigation was to examine the responses to weathering of a wide range of dwarf and semi-dwarf varieties when grown in northern N.S.W. and to assess their potential in a sprouting resistance breeding program.

MATERIAL AND METHODS

Spring wheat lines originating from Mexico, Australia or Africa, with known dwarfing gene status (Table 2), were grown in field plots (1 m x 2 row) at the Plant Breeding Institute, Narrabri in both 1980 (2 sowing times) and 1981. In addition, 4 sets of isolines for *rht*, *Rht1*, *Rht2* and *Rht1* + *Rht2* (CI 13253/7*Burt; CI 13253/7*Nord; CI 13253/7*Itana and Suweon 92/7*Burt) were obtained from R.E. Allen, Pullman, Washington, U.S.A. via the Australian Wheat Collection, Tamworth, N.S.W. and sown in field plots (1 m x 2 row) in 1981.

Estimation of resistance to weathering

150-200 heads of each line were harvested at random from field plots at harvest maturity (11-12% moisture content) and stored under cover at ambient temperature for 15 days. 30 heads were subjected to a standard weather treatment (50 mm of rain in 2 hours followed by 58 hours of high relative humidity at 20-25°C) in a controlled environment rain simulation system. Following treatment heads were dried in a forced-air-dehydrator at 35-40°C. Treated and non-treated samples were threshed and the grain reduced to flour for estimation of falling number. The extent of damage caused by the weather treatment was assessed from the difference between the falling number of weathered and non-weathered samples.

Rate of production of α-amylase

1.5 g hand-threshed grain was surface-sterilized with 1% Biogram for 10 minutes, washed 3 times with sterile distilled water, placed on filter paper in petri dishes with 4 ml sterile water containing Nystatin

(100 µg/ml) and Streptomycin (100 µg/ml), and incubated at $4°C$ for 48 hours before transferring to an incubator at $20°C$. In a parallel experiment gibberellic acid (final concentration $10^{-4}M$) was added to the petri dishes at the start of the incubation period. Duplicate samples were removed from the incubator at intervals, dried at $35-40°C$, ground and the α-amylase activity determined by the Phadebas method described by Barnes and Blakeney (1974).

RESULTS

Within each group of varieties with particular dwarfing genes or gene combinations, there was a wide range of responses to a standard weather treatment in 1981 (Figure 1). A similar pattern was noted for 2 sowings in 1980 although the response of individual varieties was not necessarily consistent from trial to trial or year to year. None of the trials were affected by rain prior to or during harvest.

4 sets of isolines also gave a wide range of responses to weathering in 1981, however, as in the previous experiment there was no apparent relationship between the response and the particular dwarfing gene(s) involved (Table 1). These isolines were not completely adapted to the local environment and seed quality was not always acceptable. In addition some lines were badly infected with disease at different stages of development.

Table 1. Effect of a standard weather treatment on 4 sets of isolines involving the Norin 10 dwarfing genes 15 days after harvest. Damage was assessed by the falling number method.

| Gene | Falling number (sec.) | | | |
	Set 1	Set 2	Set 3	Set 4
gai/rht	225	316	479	279
Gai1/Rht1	133	322	467	144
Gai2/Rht2	140	479	474	199
Gai1/Rht1 + Gai2/Rht2	77	406	330	125

Rates of production of α-amylase by grains germinated in water at $20°C$ also covered a significant range irrespective of dwarfing gene status (Table 2). When germinated in the presence of exogenous gibberellic acid ($10^{-4}M$), all varieties, with the exception of Tordo and Topo, produced approximately twice as much α-amylase. For Tordo and Topo the ratio of the amount of enzyme produced in the presence of gibberellic acid to that produced during incubation in water alone ranged from

0.9-1.3, i.e. these varieties were insensitive to the addition of exogenous gibberellic acid.

Table 2. Rate of production of α-amylase during the germination in water at $20\,^{\circ}C$ of grain from a range of dwarf and semi-dwarf wheats. Figures are the means of duplicate samples. The source of the variety is indicated by the letter in brackets (A=Australia, M=Mexico and Af=Africa).

Variety and Dwarfing Gene	α-amylase (Ug^{-1}) ÷ 10^3		
	24 hours	36 hours	48 hours
Gai1/Rht1			
(A) Condor	3.7	6.8	21.5
(A) Cook	3.2	5.5	8.5
(M) Jupateco 73	4.1	5.2	15.2
(M) Siete Cerros	2.1	5.3	18.1
(M) Penjamo 62	4.1	9.2	23.2
(Af) Dougga	1.7	3.3	13.3
(M) Lerma Rojo 64	5.7	9.7	18.3
(A) WW15	2.8	6.6	12.3
Mean	3.4	6.5	16.3
Gai2/Rht2			
(M) Pavon 76	3.0	5.0	10.0
(M) Sonora 64A	3.3	10.1	23.5
(M) Salamanca 75	2.7	4.5	7.0
(M) Pitic 62	9.3	11.6	30.3
(A) Songlen	2.6	6.2	17.0
Mean	4.2	7.5	17.6
Gai1/Rht1 + Gai2 Rht2			
(A) Shortim	1.6	2.9	10.1
(Af) Olsen	3.7	4.9	12.0
(Af) Gwebi	2.5	4.8	8.5
(M) Cajemi 71	4.5	8.3	16.2
(M) RKF 63.52	3.0	7.1	20.3
(M) RKF 63.53	4.9	8.6	26.1
Mean	3.4	6.1	15.5
Gai3/Rht3			
(M) Tordo	2.2	3.9	11.2
(M) Topo	4.1	5.9	11.2
Mean	3.2	4.9	11.2

During the 1980 season significant variation in the resistance of the variety tordo was observed. In 3 trials the response to a weather treatment 15 days after harvest was similar (falling number = 320-470 sec.) whilst in the fourth trial the level of resistance was unacceptably low (falling number = 90-180 sec.). When samples of grain from these trials were germinated in water the rate of α-amylase production in the least resistant sample was found to be similar to that in grain

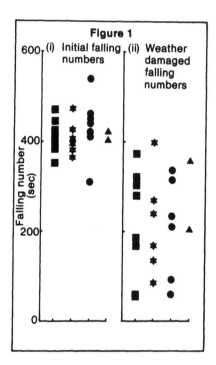 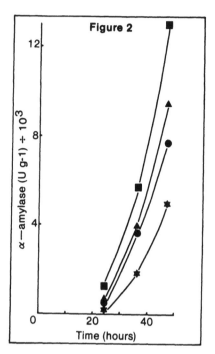

Figure 1. Distribution of falling numbers for wheats with different dwarfing genes. Falling numbers were determined before (i) and after (ii) a standard weather treatment applied 15 days after harvest maturity. $Gai1/Rht1$ (■), $Gai2/Rht2$ (✦), $Gai1/Rht1 + Gai2/Rht2$ (●) and $Gai3/Rht3$ (▲).

Figure 2. Production of α-amylase in whole grains of 2 samples of Tordo and 3 semi-dwarf wheats. Tordo $Gai3/Rht3$ - resistant sample (✦) and susceptible sample (●), Songlen $Gai2/Rht2$ (■), Cook and WW15 $Gai1/Rht1$ (▲).

of varieties such as Cook and WW15, but still less than Songlen (Figure 2). By contrast the rate of production of α-amylase in samples of grain from the more resistant plants was significantly reduced (Figure 2). For all Tordo samples, however, the ratio of the amount of enzyme produced in the presence of added gibberellic acid to that produced in the absence of exogenous hormone was similar at all the time intervals examined (ratios for the more resistant samples were 1.35, 1.3 and 1.25; and the least resistant samples 1.3, 1.5 and 1.2 for samples taken at 24, 36 and 48 hours after the start of incubation at 20^0C respectively).

DISCUSSION

Variation in both the response to weathering and
the rate of production of α-amylase by germinating grains
within the groups of wheat varieties with particular
Norin 10 genes (*Gail/Rht1* and *Gai2/Rht2*) was not unexpected. Radley (1970) and Gale and Marshall (1973)
have reported that gibberellic acid insensitivity
associated with these genes is restricted to the aerial
parts of the plant and that the endosperm or aleurone of
seeds of Norin 10 dwarfs respond to gibberellic acid in
a similar manner to seeds of tall wheats. By contrast
the occasional wide variation in the response of the
extreme dwarfs, Tordo and Topo (*Gai3/Rht3*), to weathering
was contrary to expectation. Flintham and Gale (1980)
had previously shown that the gene, *Gai3/Rht3*, repressed
flour α-amylase levels in a range of genotypes derived
from Cappelle Desprez (*Gai3/Rht3*) x Minster Dwarf (*Gai3/
Rht3*) when these lines were grown at Cambridge, U.K.
The variation in the response of Tordo at Narrabri in
1980 was closely associated with significant differences
in the capacity of the grains of the different Tordo
samples to produce α-amylase during incubation in water
at 20^0C, however, in all instances the grains retained
their insensitivity to exogenous gibberellic acid. These
observations suggest that there is a considerable environmental component in either the synthesizing potential
of the developing grain or in the time required to
initiate the synthesis of α-amylase, even in *Gai3/Rht3*
plants, and upon this system the gibberellic acid insensitivity mechanism is superimposed. Environmental
factors have been reported to influence the capacity of
grain to produce α-amylase in response to gibberellic
acid (Nicholls, 1980 and King and Gale, 1980) and the
absolute requirement in barley for gibberellic acid to
induce α-amylase synthesis (Nicholls, 1981). However,
in these investigations only distal halves of grains were
used so that data is not strictly comparable.
        The phenomenon of sprouting in the ear of wheat is
a complex process involving a number of sequential steps,
many of which are affected by the environment. The
production of α-amylase is one of the later steps in the
sequence. Whether in fact the observed response differences of and the rates of production of α-amylase in
Tordo are causally related remains to be clarified. In
view of the apparent variation in resistance and the
severe agronomic problems, the practical usefullness of
extreme dwarfs such as Tordo in north western N.S.W. must
be in grave doubt.

REFERENCES

Barnes, W.C. and Blakeney, A.B. 1974. Determination of cereal alpha-amylase using a commercially available dye-labelled substrate. Stärke 26:193-197.

Bhatt, G.M., Derera, N.F. and McMaster, G.J. 1977. Utilization of Tom Thumb source of pre-harvest sprouting tolerance in a wheat breeding programme. Euphytica 26:565-572.

Flintham, J.E. and Gale, M.D. 1980. The use of Gai3/Rht3 as a genetic base for low α-amylase wheats. Cereal Res. Commun. 8:283-295.

Gale, M.D. and Marshall, G.A. 1973. Insensitivity to gibberellin in dwarf wheats. Ann. Bot. 37:729-735.

Gale, M.D. and Marshall, G.A. 1975. The nature and genetic control of gibberellin insensitivity in dwarf wheat grain. Heredity 35:55-65.

King, R.W. and Gale, M.D. 1980. Pre-harvest assessment of potential α-amylase production. Cereal Res. Commun. 8:157-165.

Nicholls, P.B. 1979. Induction of sensitivity to gibberellic acid in developing wheat caryopses: effect of rate of desiccation. Aust. J. Plant Physiol. 6:229-240.

Nicholls, P.B. 1980. α-Amylase synthesis in the absence of applied gibberellic acid in barley endosperm halves. Proc. XIII Int. Bot. Cong., Sydney, Australia (abstr) P.62.

Radley, M. 1970. Comparison of exogenous gibberellins and response to applied gibberellin of some dwarf and tall wheat cultivars. Planta 92:292-300.

Alpha-Amylase Production in the Late Stages of Grain Development— An Early Sprouting Damage Risk Period?

M. D. Gale, J. E. Flintham and E. D. Arthur,
Plant Breeding Institute, Cambridge, England

INTRODUCTION

The propensity of some wheat varieties to have unacceptably high levels of α-amylase at maturity even in the absence of visible sprouting has been recognised for some years. Bingham and Whitmore (1966) considered the problem in some British varieties to be due to high residual levels of the enzyme present in developing grains. They concluded that the syndrome was relatively simply inherited, and could thus be avoided in the future breeding of bread-making quality wheats.

More recent evidence has suggested that this source of grain α-amylase may be more widespread than previously thought. As shown below, the enzyme in the Plant Breeding Institute varieties, such as Maris Huntsman, Mardler and Norman, which all have high levels at maturity, is not the form found during grain development, α-AMY2, but the malt α-amylase α-AMY1. This is the form of α-amylase potentially more deleterious in the bread-making process (Sargeant and Walker 1978) and must be being produced 'de novo' by grain mobilization processes. α-AMY1 has been shown to be present before grain maturity in the French wheat, Champlein (Sargeant 1980) and in Canadian varieties (Marchylo, LaCroix and Kruger 1980).

This paper describes experiments designed to ascertain how important prematurity production is as a component of harvest α-amylase levels. In particular a) how general is expression of the character among wheat varieties, b) how much genetic variation for expression of the trait is available, c) what environmental factors regulate the degree of 'early' α-AMY1 production and, most importantly, d) what are the consequences of 'early' grain mobilization on α-amylase accumulation and preharvest sprouting during the more conventional 'late' risk period.

RESULTS AND DISCUSSION

A first experiment was conducted with two spring varieties,
Snabbe and Aotea, in controlled environments. The Swedish variety
was chosen as an example of a relatively sprouting-resistant genotype and the white-grained New Zealand variety for its susceptibility. In neither variety has prematurity α-amylase production been
previously reported. The grains were sampled at three-day intervals
from ten days post anthesis (dpa) to about 50 dpa, and then less
frequently until 100 dpa. Each sample was tested for embryo germinability, whole grain germinability, distal half grain α-amylase production after incubation with gibberellic acid (GA) and endogenous
α-amylase content. The plants were grown in two cabinets, both at

FIGURE 1

Grain development, germinability and α-amylase content in the variety Snabbe grown in 'fast' and 'slow' grain drying environments.

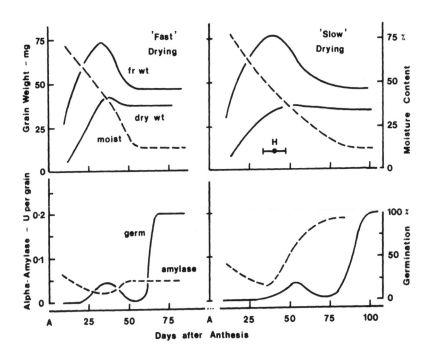

Note. All grain sampled from 1st and 2nd florets of central spikelets of main shoots. Germinability measured as % grains
with ruptured seed coat after 7d at 23°C on moist filter
paper. α-Amylase levels measured as described by Gale and
Marshall (1975) and calibrated using the Phadebas (Pharmacia
AB, Uppsala) test.

15°C for a 20 h light period and 10°C for a 4 h dark period. In the
'fast' drying treatment humidity was maintained at 56% r.h. In the
'slow' drying treatment, r.h. was boosted to about 92% by injecting
a fine spray of water into the air flow for 10 min every hour. This
treatment was commenced just before the grains attained maximum
fresh weight ('H' in Figure 1).

Firstly the results showed that both varieties exhibit a significant increase in α-amylase levels during the grain drying period.
The increase was least marked in Snabbe (Figure 1) but was especially clear in the slow drying treatment. Once the maximum level
was attained, when the grains had reached about 20 per cent moisture
content, extractable enzyme activities remained constant for the
duration of the experiment. For reference, the 'fast' drying treatment produced grain development and ripening rates of the same order
as an 'average' British growing season.

No visible germination in the ear was observed in the experiment. However, detached grains of both varieties showed some limited capacity for germination during the ripening period. The separate measurements of embryo germinability and distal half grain responsiveness to GA showed that a short period of limited whole grain
germinability coincided with a similar 'window' of aleurone responsiveness. The capacity of the embryo to germinate when separated
from the endosperm appeared not to be involved. Embryo germinability on agar was 100% and was attained at about 30 dpa and was maintained at this level, even through the period of whole grain and
aleurone dormancy.

Clearly α-amylase production before maturity may be a general
phenomenon which appears to be enhanced under slow grain drying conditions. In addition, the occurrence of limited detached grain germinability and aleurone GA responsiveness during approximately the
same period confirms that the enzyme production is related to grain
mobilization processes.

The nature of α-amylase produced during the early period is
shown in Figure 2. In a similar controlled environment experiment
Maris Huntsman grains, sampled a week before harvest ripeness,
showed variable levels of α-AMY1, whereas earlier in development
only α-AMY2 had been present. A final sample demonstrates that
similar levels of α-AMY1 are retained in a similar proportion of the
grains long after maturity in the absence of visible preharvest germination.

A similar experiment to that described above was carried out in
the field in 1981. Several varieties were sampled at weekly intervals for enzyme content, germinability and aleurone GA-sensitivity
and harvest was delayed to increase the chance of observing visible
sprouting damage. The enzyme levels in four of the genotypes are
shown in Figure 3. These confirm that varietal variation exists for

the level of enzyme produced before dormancy. Bezostaya I is unrelated to, but has just as high α-amylase as Maris Huntsman.

FIGURE 2

α-Amylase isozymes in typical single grains of Maris Huntsman, sampled during grain development and several weeks after ripeness.

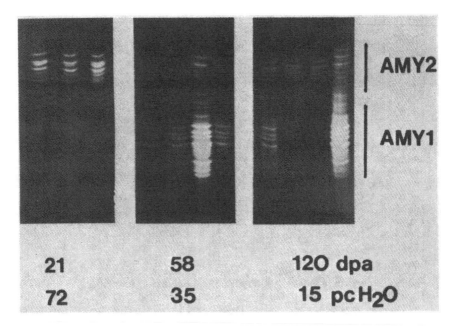

Note. Growth conditions 18 h light at 15°C, 85% r.h., 6 h dark at 10°C. Harvest ripeness at 64 dpa. Isoelectric focussing as described by Sargeant and Walker (1978).

Two other points, relating to the initial aims of these experiments, are indicated from these varietal comparisons. First, the propensity to produce higher levels of α-amylase before maturity is unrelated to the duration of dormancy. Maris Huntsman and Bezostaya I have longer dormancy periods than the two lower α-amylase varieties in Figure 3. RL4137, the genotype with longest dormancy in the experiment, had maturity enzyme levels midway between the two pairs of varieties shown.

Secondly, and more significantly, once sprouting proper has begun the high α-amylase varieties suffer most damage both in terms of enzyme levels attained and observed germination in the ear. It appears as if grain mobilisation initiated during the drying period is merely halted during dormancy and continues with the characteristic logarithmic increase in α-amylase level when pre-harvest sprouting occurs in the 'late' risk period.

FIGURE 3

α-Amylase in grains
sampled from field
grown plots of four
wheat varieties.

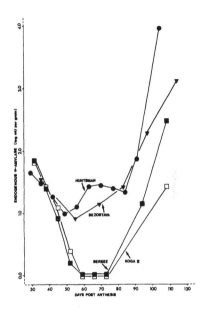

Note. Spikes were sampled from main tillers and fast frozen prior to freeze drying. Visible germination at final harvest, Maris Huntsman 16%, Bezostaya I 12%, Bersée 4% and Koga II 1%.

Finally, a comparison of the α-amylase levels in field and growth cabinet grown samples of Maris Huntsman grain provides confirmation of the effect of drying rate on early enzyme accumulation (Figure 4).

CONCLUSIONS

The simplest model to fit these results involves three factors, i) the attainment of aleurone (and scutellum) competency to respond to GA during ripening, ii) the availability of a promoter in the grain at the same time (possibly a net excess of GA over abscisic acid), and iii) the availability of adequate grain moisture to allow the production of hydrolases to proceed.

The involvement of GA mediated aleurone and/or scutellum response is indicated both by the nature of the α-amylase isozymes produced during the 'early' risk period and by the fact that the only genotypes in which we have observed no prematurity enzyme production are Minister Dwarf and another Tom Thumb dwarf line, D6899. The Tom

FIGURE 4.

Endogenous α-amylase levels and drying curves of Maris Huntsman grain in controlled (o) and field (•) environments.

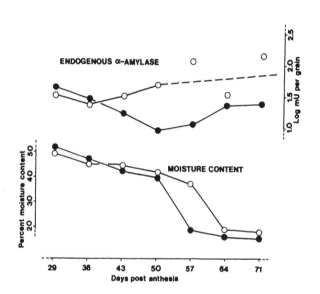

Note. At 71 dpa grains were still completely dormant in both environments, no visible sprouting was observed at this time. Grain growth curves and final grain weights were similar in both environments.

Thumb dwarfing gene, Rht3, renders aleurone (and scutellar) tissue relatively insensitive to GA in terms of α-amylase production (Gale and Marshall 1973, 1975).

It is not clear whether the enhancing effect of slow drying is due simply to an extension of the duration of the risk period, or to a direct effect of drying rate itself. King and Gale (1980) reported exactly such an effect of drying rate on the potential of detached distal half grain to respond to GA. It seems likely now that the aleurone responses monitored in those experiments were, in fact, those during the early sprouting damage risk period.

It is evident that two of the three factors in our model could be under genetic control, i.e. available hormone levels and degree of aleurone response. The duration of grain drying would appear to be environmentally determined.

The significance of this source of sprouting damage is still not clear. However the data shown in Figure 3 indicate that early grain mobilization may increase the risk of, and damage during, post-dormancy sprouting. This may be linked with the fact that in 1977, when the most severe sprouting damage of the last decade was experienced in the South of England (Home Grown Cereals Authority, 1977), the rate of water loss during grain maturation was particularly slow (Mitchell, Black and Chapman 1980).

Of course, this causal relationship between 'early' and 'late' enzyme production, inferred from comparisons of a few varieties in the experiments described here, could be coincidental. Further studies to confirm the result in more appropriate genetically defined genotypes are underway.

It is imperative that the relative importance of early and late sprouting damage is resolved because many breeders are employing tests based on post ripeness sprouting scores in artificial rain chambers such as that described by McMaster and Derera (1976). These breeders' screens will only effectively monitor the duration of dormancy and could therefore be ignoring damage incurred before the tests are conducted.

REFERENCES

Bingham, J. and Whitmore, E.T. 1966. Varietal differences in wheat in resistance to germination in the ear and α-amylase content of the grain. J. Agric. Sci. 66:197-201.

Gale, M.D. and Marshall, G.A. 1973. Insensitivity to gibberellin in dwarf wheats. Ann. Bot. 37:729-735.

Gale, M.D. and Marshall, G.A. 1975. The nature and genetic control of gibberellin insensitivity in dwarf wheat grain. Heredity 35:55-65.

Home Grown Cereals Authority. 1977. Cereal quality survey, Hamlyn, London.

King, R.W. and Gale, M.D. 1980. Preharvest assessment of potential α-amylase production. Cereal Res. Commun. 8:157-166.

McMaster, G.J. and Derera, N.F. 1976. Methodology and sample preparation when screening for sprouting damage in cereals. Cereal Res. Commun. 4:251-254.

Marchylo, B.A., LaCroix, L.J. and Kruger, J.E. 1980. α-Amylase isoenzymes in Canadian wheat cultivars during kernel growth and maturation. Can. J. Plant Sci. 60:433-443.

Mitchell, B., Black, M. and Chapman, J. 1980. Observations on the validity of predictive methods with regard to sprouting. Cereal Res. Commun. 8:239-244.

Sargeant, J.G. 1980. α-Amylase isoenzymes and starch degradation. Cereal Res. Commun. 8:77-86.

Sargeant, J.G. and Walker, T.S. 1978. Adsorption of wheat alpha-amylase isoenzymes to wheat starch. Staerke 30:160-163.

Ear Wetting and Pre-Harvest Sprouting of Wheat

R. W. King and H. Chadim, *CSIRO, Division of Plant Industry, Canberra, Australia*

INTRODUCTION

Little attention has been paid to the effect of ear and grain wetting rates on pre-harvest sprouting of wheat. Water infiltration rates into mature grain of wheat do differ between cultivars (Stenvert and Kingswood, 1976; Clarke, 1980). The structure of the ear may also be important but present evidence is inconclusive. For example, it has been reported that awns in wheat have no effect (Clarke, 1982) or hasten (Pool and Patterson, 1958) the drying of ears after wetting. In barley ear nodding angles of greater than 120° from the upright reduce water damage (Brinkman and Luk,1979) but it is uncommon for wheat ears to nod at these angles. However, in relation to the question of pre-harvest sprouting the major shortcoming of these studies is the lack of information on germination itself.

In the present investigation with wheat we have examined a) the relationship between rainfall, ear and grain wetting and germination and b) the nature of physiological and morphological controls over ear wetting and sprouting. These results have been presented in greater detail elsewhere (King 1983a).

MATERIALS AND METHODS

The fifty or so cultivars examined were mostly white grained early and modern Australian spring wheats as well as some awned/awnless near-isogenic research lines developed by Dr. R. Richards (CSIRO). Apart from the near-isogenic lines all the ears came from material grown in the field at the same site and time. All ears were held dry in the laboratory for at least six months after harvest so that serious differences in post harvest dormancy were avoided and so that the grain equilibrated to a similar low moisture content.

Water uptake by four replicate ears was followed over 30 h for ears fixed vertically in a spray misting chamber which gave the equivalent of 62 mm day^{-1} rainfall. Ears were also placed horizontally in one experiment.

Germination of grain in the ear was scored at the end of the 30h wetting treatment. For isolated grain, measurements were also made of water uptake from water saturated filter paper and of germination over 30h.

Water uptake by ears and isolated grain has been expressed as a percentage to allow for varietal differences in weights of ears and grains. The % uptake at any time is:

$$\text{Fresh weight - initial air dry weight} \times 100 \over \text{fresh weight}$$

RESULTS AND DISCUSSION

Ear Wetting and Sprouting

Exposure of ears to spray misting equivalent to 62 mm day^{-1} rainfall resulted in a reasonably rapid initial water uptake. However, by 6h, ear water uptake had virtually plateaued at about 80% of that at 30h (Fig. 1).

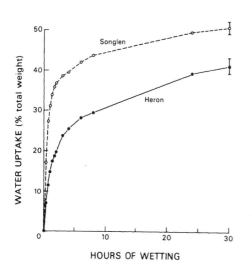

FIG. 1. Time course of ear water uptake (%) for air dry ears of cvs Songlen and Heron. Rainfall at a simulated rate of 62 mm day^{-1}. Standard error 4-8% of measured values (2 x SE shown).

Varieties ranged widely in both initial (2h) and plateau (30h) values of ear water uptake. However, rankings at each time were relatively fixed at least for the extreme cultivars. Moreover, for all 51 cultivars regression analysis of ear water uptake rate at 2h versus 30h indicated a highly significant correlation (r = 0.79). Low initial uptake was associated with reduced uptake at 30h and so forth.

Since all ears had grown under similar conditions it is unlikely that there were environmental differences between varieties. Effects attributable to inherent differences in ear size are reduced by calculating relative (percent) water uptake. Also, within one cultivar

(Egret) water uptake values were relatively constant for ears analysed at three separate times over seven months. (Average 2h uptake 30.6 ± 1.0%). Therefore, the spread of cultivar ear wetting rates points to adequate genotypic differences for selection by plant breeders.

Ear Characteristics and Water Uptake
        Stage of Ear Development   Fresh tissue absorbs little water but as the ear and grain dehydrates at maturity there is a three - to five-fold increase in ear wetting (King, 1983a). The timing of this change has obvious implications for field studies of rain damage but cannot account for the varietal differences seen (King, 1983a) for material harvested dry and kept for at least six months prior to rewetting.
        Ear Angle   Reduction of rain damage in barley has been correlated with nodding of the ears at angles greater than 120° from the vertical (Brinkman and Luk, 1979). Presumably water uptake was reduced. By contrast, for wheat ear-nodding angles generally do not exceed 90° from the vertical. Furthermore, from tests with two varieties, ears held horizontally absorb water more rapidly at the start of wetting (Table 1). This more rapid absorption of water also correlates with a slight but non-significant stimulation of germination over the 30h period.

TABLE 1.   Effect of ear angle on water uptake after 2h and sprouting after 30h for two wheat cultivars exposed for 30h to simulated rainfall. Data for cultivars combined as interactions were not significant. Significance indicated as *** P = 0.001.

Ear angle	2h water uptake (%)	30h germination (%)
Vertical	29.8	61
Horizontal	34.5***	71^{ns}

        Effect of Awns   The presence of awns hastens water loss from wheat ears during drying and could enhance uptake during wetting (Pool and Patterson, 1958). Therefore, ear wetting was examined in near-isogenic lines selected for the presence or absence of awns. This study is to be presented in more detail elsewhere but preliminary measurements show significantly slower ear water uptake in the absence of awns (Table 2). Germination in the ear was reduced but not significantly, in the non-awned lines. However, grain isolated and imbibed on filter paper showed no differences between awned and awnless genotypes in their germination or water uptake.

TABLE 2 Effect of the presence or absence of awns on 2h and 30h water uptake and on germination in the ear at 30h for near-isogenic lines of wheat derived from a cross Condor x Russian. Germination at 30h of isolated grain is for grain taken from comparable ears. Significance indicated as *P = 0.05 .

Genotype	Ear Water Uptake		In-ear 30h	Isolated Grain
	2h (%)	30h	Germination (%)	Germination (%)
Awned	36.8±1.1	55.3±2.8	38.7±2.7	41.0
Awnless	27.9±0.9*	50.7±1.2^{ns}	33.5±3.2^{ns}	41.0

Varietal Relationships between Awns, Ear Wetting and Sprouting Since the awned character modifies ear water uptake (Table 2) the effect of awns was examined in the fifty one varieties examined for water uptake and in-ear sprouting. As shown in Table 3 ears of awned varieties absorbed water more rapidly and their grain germinated better. Both effects were statistically significant and the changes parallel those obtained with lines near-isogenic for the presence and absence of awns.

TABLE 3 Effect of the presence or absence of awns on water uptake and germination of wheat in the ear. Significance indicated as: *P = 0.05; **P = 0.01; ***P = 0.001 (from King 1983a)

Character	Ear Water Uptake		In-Ear 30h
	2h (%)	30h	Germination (*)
Awned (26 cultivars)	31.8±0.7	48.1±0.7	41.8±5.0
Awnless (25 cultivars)	27.1±0.8***	45.6±0.7*	17.2±4.0***

Despite the consistent depression of germination in the ear of the slower wetting, awnless lines (Table 2,3)

causality cannot be assumed for varietal comparisons.
There could be a totally spurious varietal grouping of
dormancy differences which would, by chance, give an
apparent relationship. However, by measuring germination
of isolated grain simultaneously with in-ear germination
it was possible to make allowance for varietal germination
differences. Figure 2 illustrates this approach.

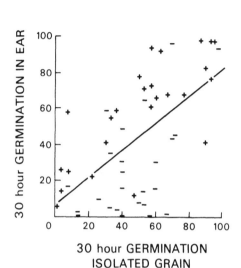

FIG. 2. Relationship
between germination
in the ear and of
isolated grain on
filter paper.
Designation of
varieties as awned
(+) or awn less (-).
From King (1983a).

30 hour GERMINATION
ISOLATED GRAIN

As expected, germination in the ear is significantly
correlated with simultaneous, 30h germination of isolated
grain on filter paper (r = 0.61). However, superimposed
over this correlation, there is a distinctive clustering
of awned and awnless varieties. Seed from awnless
varieties germinated relatively poorly in the ear compared
with that on filter paper. For the awned varieties
germination was on the whole better in the ear than on
filter paper. Clearly the ear and, more specifically the
awns influence grain germination. Thus, for varieties,
as well as for isogenic lines there appears to be a real,
causal relationship between possession of awns, ear water
uptake (Table 2) and in-ear sprouting (Figure 2).

Grain Characteristics and Sprouting
        It is well known that smaller wheat grains germinate
more rapidly (Bremner pers. comm.). This explains why
germination in the ear was always more advanced in the
smaller grain in peripheral floret positions. Figure 3
illustrates this observation for in-ear germination of the
cultivar Kalkee following 30h of misting. For all fifty
one cultivars grain removed from floret positions three
and above also germinated significantly better than that

from positions one and two when germinated on filter
paper (King, 1983b). Clearly, this effect of floret
position on germination in the ear is inherent in the
grain and not related to ear wetting characteristics.

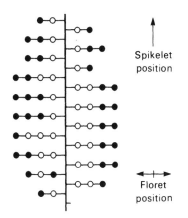

Spikelet
position

FIG. 3. Germination in
the ear (●) or non-germinat-
ion (0) of cv. Kalkee
determined after 30h of
simulated rain.

Floret
position

CONCLUSIONS

Based on the foregoing findings, a wheat variety bred
for reduced pre-harvest sprouting should have no awns.
Considering ear wetting, alone, other factors of importance
might include upright ears and ears which take a prolonged
period to dry down at maturity. Varietal differences in
grain water uptake might also be examined and we have
reported elsewhere on differences between the cultivars
examined in these experiments (King, 1983b).

Ultimately, it is changes in grain dormancy which
will have greatest impact on the control of sprouting.
The range of germination seen on filter paper at 30h
(Fig. 2) emphasises this point. Nevertheless, changing
ear structure to alter ear water absorption will have
none of the side effects on grain quality and yield that
might arise when grain characters are manipulated.
However, reduced sprouting in awnless varieties may only
follow when the delay in ear wetting is a significant
proportion of the time taken for germination (cf. varietal
differences in Figure 1). Obviously, the effect of awns
will be even greater in reduced rainfall. On the other
hand, in a more dormant variety slower germination might
obliterate any effects via awns and wetting time. The
value of the awnless character is also uncertain when
rain falls as brief showers. If conditions after a rain
shower favour slow evaporation then the wetting response
should compare to continuous rainfall. However, ears
which wet slowly should also dry more slowly (Pool and
Patterson, 1958). Thus, under intermittent rainfall,
the faster wetting of an awned variety may not be a

drawback in terms of total duration of wetting. Computer modelling may be required to define these complex relationships between rainfall input, the rates of ear wetting and the rate of germination.

Another component of sprouting that warrants study is the effect of grain size and of grain number per spikelet on germination in the ear. The smaller grain of the peripheral floret positions germinates most rapidly so that selection for larger grain may be of value. Selection for ears with fewer spikelets but more grain per spikelet might also reduce the proportion of sprouted grain since the two or three most peripheral grain germinate first (e.g. Figure 3 and King 1983b). The major drawback to either of these suggestions could be adverse effects on yield. Also, if yield increase comes with reduced grain size then care should be taken to avoid associated problems of sprouting.

ACKNOWLEDGEMENT
Dr. R. Richards is particularly thanked for providing the near-isogenic lines.

REFERENCES

Brinkman, M.A. and Luk, T.M., 1979. Relationship of spike nodding angle and kernel brightness under simulated rainfall in barley. Can. J. Pl. Sci. 59, 481-485.

Clarke, J.M., 1980. Measurement of relative water uptake rates of wheat seeds using agar media. Can. J. Pl. Sci. 60, 1035-1038.

Clarke, J.M., 1982. Effect of awns on drying rate of windrowed and standing wheat. Can. J. Pl. Sci. 62, 1-4.

King, R.W., 1983(a). Water uptake and pre-harvest sprouting damage in wheat; ear characters. Aust. J. Agric. Res. (submitted).

King, R.W. 1983(b). Water uptake and pre-harvest sprouting damage in wheat; grain characterstics. In preparation.

Pool, M. and Patterson, F.L., 1958b. Moisture relations in soft red winter wheats. II Awned versus awnless and waxy versus non-waxy glumes. Agron. J. 50, 158-160.

Stenvert, N.L. and Kingswood, K. 1976. An autoradiographic demonstration of the penetration of water into wheat during tempering. Cereal Chem. 53, 141-149.

Temperature-Moisture Interactions as a Factor in Germination and Dormancy

W. Woodbury and T. J. Wiebe, *University of Manitoba, Winnipeg, Manitoba, Canada*

SUMMARY

Experiments were carried out to investigate effects of temperature on germination of wheat under conditions where rate of water
uptake was restricted by resistances outside the seed. At low temperature, the lag in germination or emergence is increased but there
is little effect on the slope of the growth curve. Temperature
altered the amount of moisture required for germination; at low temperature, the amount was decreased if water supply was interrupted
but it was increased if water supply was continuous. Winter wheat
required less water to germinate. Results are discussed in terms of
possible pathways for water movement within the seed.

INTRODUCTION

Our interest in this area originated with a question from L.E.
Evans (Department of Plant Science, University of Manitoba), "would
it be possible to use a germination inhibitor to permit fall planting of cereals?" In many areas of the Canadian prairie, winter
wheat does not survive because of very low temperatures in the
absence of adequate snow cover. With spring crops, seeding is often
delayed because of very wet soil conditions. As the frost-free
growing season is not more than 120 days, early emergence of fall-
planted, treated seed could increase yield given a longer season and
because grain filling might occur earlier and before temperature and
moisture stresses developed.

In this area, surface soil temperatures would be below 5°C
between October 15 and April 15. Dormancy of fall-planted seed
would be at low temperature, and germination would occur at 5-10°C.
Soil moisture, while variable, would be at water potentials below
the level (near 0 bars) of wet filter paper systems. Hydraulic conductivity of soil would be much lower.

The seed germination literature provides us with little more
than the cardinal temperatures but there is interest in low temperature germination of heat-loving crops (Khan 1977). Most work with
inhibitors and promoters of germination has been done at 20°C and on
wet filter paper. The literature on seed ecology may provide us
with some clues. Figure 1 presents idealized diagrams for three

44

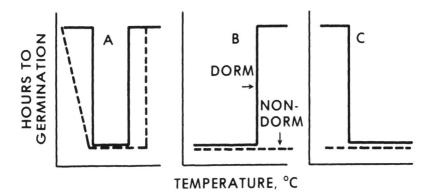

TEMPERATURE, °C

Fig. 1 Idealized diagrams for three patterns of thermal dormancy.

patterns of thermal dormancy. Diagram A shows seeds which, at maturity, are dormant at temperatures above and below some definite range (Thompson 1970). Upon loss of dormancy, the permissible temperature range widens through a shift in one or both boundaries. Diagram B shows the response seen in wheat, barley and oats (MacKey 1976); dormancy is only expressed at higher temperatures. In terms of fall-planting, it is a type-C response that we would hope to be able to impose since in spring dormancy would disappear in warmer soils. In many species, the boundaries are very sharp; a shift of a few degrees converts the seed from the fully dormant to the nondormant condition. This is usually interpreted in terms of effects of temperature on the phase properties of membranes. However, a case can be made for critical temperature effects on the phase properties of structural polymers within the seed.

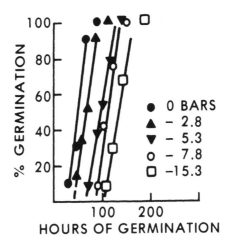

Fig. 2. Germination of wheat in soils with different water potentials at 20°C.

Figure 2 shows the germination curves for wheat in soil at 20°C where water potential (ψ) was controlled through equilibration against polyethylene glycol solutions (Blackshaw et al 1981). Lower ψ increased the lag period before germination began but had little effect on the slope of the germination curve. What is more surprising is that low temperature produced the same result for germination and for seedling emergence (Blackshaw et al 1981; Hallem 1981; Khan 1977). Evidently the seeds must undergo physiological adaptation during the lag period if growth and metabolism are to be independent of temperature and ψ. Possibly this is the result of selection for uniformity of crop stand, but germination of green foxtail (Setaria-viridis) showed the same response.

The fact that the germination curves are steep complicates experiments on germination since it is difficult to obtain enough data points within 30-40 hours. In addition, high variability is expected due to small differences in duration of the lag period characteristic of individual seeds. We have taken some pains to evaluate and control possible sources of variation in our experiments.

METHODS AND RESULTS

In most of our experiments, acid washed quartz sand (Fisher Scientific; for water analysis) of a narrow range of particle sizes (85-125 mesh) was used. Water (4 to 16 mL was mixed with 65 g of the sand and the mixture compacted with a 500 g weight into plastic containers. The ψ ranged from 0 bars down to about -0.1 bars. The hydraulic conductivity of the sand was not measured, but because of the relatively large particle size, it would decline very steeply with decreasing moisture content. The driest sand had ten times more water than required for the complete germination of the seed sample (50 seeds). Thus, amount of water was not limiting but rate of water uptake was limited by resistances external to the seed with less than 14 mL of water present. Control of temperature was by means of germination cabinets or a temperature gradient table.

Seed size is an obvious source of variation. Seeds were sized on standard screens. In addition, medium-sized seeds were further separated on the basis of seed density using an air-flow seed cleaner. The seeds were set to germinate at 15°C in sand containing 10 mL of water. The results (Fig. 3) indicate that smaller seeds reach 50% germination (G/2) sooner than larger seeds. Also denser seeds take longer to germinate. Density differences could result from compositional differences of seed reserves or from structural differences. Either could affect water movement within the seed (Campbell 1958; Stenvert and Kingswood 1976).

When seed size was considered in relation to the range of sand moisture, the results were as in Figure 4. There was an abrupt increase in germination time below 16 mL of water. Under drier conditions, moisture level had relatively little effect but germination

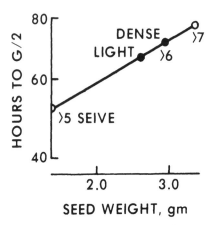

Fig. 3. Effect of seed weight and seed density on time required to reach 50% germination (G/2). Sieve number.

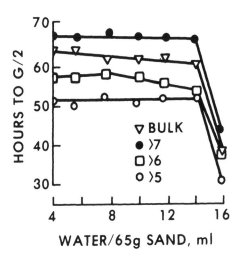

Fig. 4. Influence of sand moisture content and seed size on time required to reach 50% germination.

was affected by seed size. At moisture levels above 14 mL, a thin layer of liquid water on the surface permitted wetting of a considerable area of the seed coat, resulting in a low contact resistance to water movement. The thin film of surface water under high moisture conditions corresponds to conditions in the more usual petri dish-filter paper experiment. In most of the experiments to follow, medium-sized seeds were used.

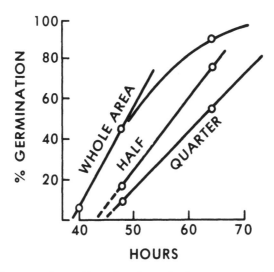

Fig. 5. Influence of seeding density of wheat on germination; 1/4, 1/2 or the entire sand area seeded.

The effect of seeding density on germination of medium-sized wheat kernels is shown in Figure 5. Fifty seeds were spread over different fractions of the moist (10 mL water) sand surface. The decrease in germination rate on smaller areas could have two origins. Initial water uptake would dry the adjacent sand and reduce its hydraulic conductivity markedly, which would slow water uptake and germination. Alternatively, inhibitors may be present in the seed which are effective under these conditions but would not be detected under conditions where the inhibitors could diffuse freely into liquid water external to the seed; i.e., movement of water toward the seed through the moist sand could prevent outward diffusion of inhibitors. Both effects would be greater when seeds are crowded together.

Effects of temperature on germination and water uptake by wheat seeds are shown in Figs. 6 and 7. Different results were obtained depending on how the experiment was done. In some experiments, large numbers of samples were prepared and samples were sacrificed at intervals for determination of germination and moisture (solid lines). Under these conditions germination was slower at low temperatures (Fig. 6) and moisture content increased (Fig. 7). This could mean either that at low temperature, more time was available for water uptake, or that more moisture was required for germination. Sorption theory (Mohsenin, 1970) predicts a higher equilibrium moisture content at low temperature. Norstar (winter) wheat germinated faster and at lower moisture than did Neepawa (spring) wheat. If seed was returned to the containers after counting germination and weighing at each interval, the curve for germination was similar, however, now moisture content increases with temperature

48

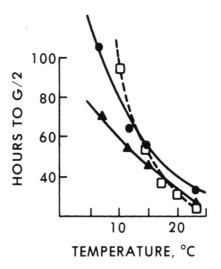

Fig. 6. Effect of temperature on time required to reach 50% germination: —●— Neepawa spring wheat; —▲— Norstar winter wheat;
—□—Neepawa seeds returned to the container.

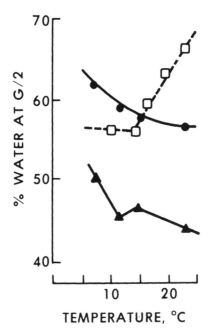

Fig. 7. Effect of temperature on moisture content when wheat
achieved 50% germination. Symbols as in Fig. 6.

(broken lines). Only the data for Neepawa wheat are presented but again, Norstar germinated sooner and at lower moisture content than did Neepawa. It is not known whether the difference is varietal or is related to the spring-winter habits.

The contrasting effects of procedure on moisture content may reflect differences in movement of water within the seed. Where the seeds are in continuous contact with the moist sand, one pathway for water movement would be established and maintained whereas with intermittant contact, a new path could be established each time the seed is returned to the container. Experiments in the literature have been done in both ways but an effect on moisture has not been reported previously. Seeds of some species remain dormant unless exposed to diurnal cycles of moisture and temperature (Khan, 1977). There was no detectable water loss during the weighing period in our experiments but it is possible that the effect is somehow related.

Norstar wheat germinated more rapidly than Neepawa, especially at the lower temperature, and at considerably lower moisture level under both experimental conditions. Presently these differences seem to be variety related although it is not certain that the winter-spring difference is involved.

DISCUSSION

The moist sand system was designed to impose resistances to water movement in regions outside the seed itself, i.e., in the sand and at the contact between the seed and the sand. Moisture and ger-minability are quite different under these conditions than when the seed contacts liquid water. Becker and Salans (1956) showed that at low seed moisture, water moved as thin "films" of bound water. This movement would be quite sensitive to temperature since the bound water has a higher activation energy than does free water. At higher moisture levels, above about 20%, the adsorbing surfaces are saturated and liquid water is held in microcapillaries of the seed structure. Movement at high moisture content should be relatively insensitive to temperature.

We believe that where external resistances limit water uptake, the initial movement within the seed would tend to be in the film mode and distribution would tend to equilibrate throughout the seed. Above the 20% level, water movement would tend to be as capillary (free) water. At about 20% moisture, the seed structure changes from an inelastic structure with limited ability to accommodate additional water to a more plastic structure which could swell quite freely (Campbell and Jones 1955; Mohsenin 1970). A number of papers (Khan 1977) have shown that if the moisture content of seed of high temperature crops is brought slowly and at high temperature to about 18%, the seeds are able to germinate quite vigorously at tempera-tures below 10°C, whereas seeds not so conditioned will be damaged or killed at the same temperature. The structure of the treated seed would be plastic at this moisture content and could accommodate the swelling induced by the greater amounts of water which are

absorbed at low temperature. Low temperature hydration of untreated
dry seed would result in non-uniform distribution of water. The
mechanical pressures which develop in the hydrated and swollen portions of the seed could result in fracturing of tissues and reduced
viability at low temperature. In the past, temperature effects on
germination and dormancy have been interpreted in terms of membranes
and inhibitor-promoter interactions. However, membranes are not the
only biological structures which show phase changes at critical temperatures. Most polymers also display this property. All previous
work has been done under conditions where the seed contacted liquid
water and usually at temperatures about 20°C. Perhaps there is more
to the story.

REFERENCES

Becker, H.A. and Salans, H.R. 1956. A study of the desorption iso-
therms of wheat at 25 and 50°C. Cereal Chem. 33:79-91.

Blackshaw, R.E., Stobbe, E.H., Shaykewich, C.F. and Woodbury, W.
1981. Influence of soil temperature on green foxtail estab-
lishment in wheat. J. Weed Sci. Soc. Amer. 29:179-184.

Campbell, J.D. 1958. Effect of mechanical damage to wheat grains
during scouring on their subsequent absorption of water during
washing. Cereal Chem. 35:47-56.

Campbell, J.D. and Jones, C.R. 1955. Effect of temperature on the
rate of penetration of moisture within damped wheat grains.
Cereal Chem. 32:132-139.

Hallem, P.M. 1981. Effects of soil temperature on root development
in wheat. M.Sc. Thesis, U. of Manitoba.

Khan, A.A. 1977. Physiology and biochemistry of seed dormancy and
germination. North Holland Pub. Amsterdam.

MacKey, J. 1976. Seed dormancy in nature and agriculture. Cereal
Res. Commun. 4:83-91.

Mohsenin, N.N. 1970. Physical properties of plant and animal materials. Gordon & Breach Publishing, New York.

Stenvert, N.L. and Kingswood, K. 1976. An autoradiographic demonstration of the penetration of water into wheat during tempering. Cereal Chem. 53:141-149.

Thompson, P.A. 1970. Germination of species of Caryophyllaeae in
relation to their geographic distribution in Europe. Ann. Bot.
34:427-449.

A Possible Role for the Pericarp in Control of Germination and Dormancy of Wheat

W. Woodbury and T. J. Wiebe, *University of Manitoba, Winnipeg, Manitoba, Canada*

SUMMARY

Wheat was germinated at 15°C on sand of varying moisture content. Coumarin only inhibited germination when the seeds contacted
liquid water. Full expression of dormancy (sprouting resistance)
was only seen under these conditions. A crack in the pericarp within the brush region, appears to be formed during seed maturation.
It allows rapid movement of dye solution beneath the pericarp in the
dorsal surface. Water movement through this pathway may transfer
inhibitors to the embryo.

Effects of temperature on interactions between water and polymers are briefly reviewed. It is suggested that temperature and
rate of drying would determine the final conformation of polymers
within seed structures thereby determining physical properties on
rehydration. These effects could be the basis of variation in germinability and dormancy.

INTRODUCTION

It is well known that environmental conditions during grain
maturation and drying can influence germinability and dormancy of
the ripe grain. Stoy and Sundin (1976) looked at grain moisture,
germinability and amylase production in response to gibberellic acid
(GA) over a period of three months following maturity. In three
wheat varieties, the initial moisture loss was accompanied by an
increase in germinability and GA response. Over the next few
months, grain moisture oscillated with environmental conditions between about 15 and 30 percent. Germination and GA response shifted
in synchrony with the moisture. Interactions between GA and ABA or
catechin-tannin in respect to germination of isolated embryo also
oscillated. The fact that moisture content of the harvested grain
influenced the various responses in assays which involved incubation
for several days under conditions where water should have been
freely available is intriguing. King and Gale (1980) confirmed the
earlier finding of Nicholls (1979) that rate of drying of detached
grain influenced the responsiveness of the grain to GA; slow-drying
was found to give highest reponse. Subsequently, Nicholls (1980)
found that temperature of drying was also involved. Grain dried at
about 20°C produced α-amylase whether or not GA was included in the

assay. Grain dried at lower temperatures required GA. Mitchell et al (1980) reported that germinability of wheat began well before maturity and natural loss of water from the grain. However, the pericarp began to lose water from about ten days after anthesis and lost more than half its water by the time germinability began to rise at about day 30. Earlier, Wellington and Durham (1961) suggested that differences in the mechanical properties of the pericarp of white and red wheats might control germinability. Belderok (1976) found that the testa of a sprouting resistant wheat maintained its integrity beyond grain maturity whereas the structure became granular before maturity in wheats without sprouting resistance.

METHODS AND RESULTS

Mixtures of quartz sand and water were used as a germination medium, the object being to impose limits to seed water uptake in regions outside the seed as would occur in a natural soil. Hydraulic conductivity of the medium, and not water potential (ψ), was the limiting factor.

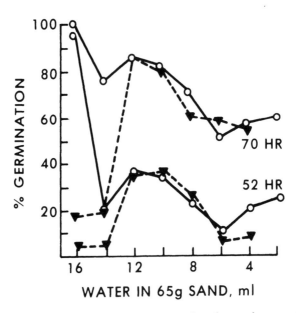

Fig. 1. The influence of sand moisture level on the germination of untreated (O) and coumarin-treated (▼) wheat after 52 and 70 hours of germination.

The progression of germination at various moisture levels in untreated seed is shown in Figure 1. Germination measured after 52 and 70 hours is very rapid at the high moisture end, but is delayed under drier conditions. Wheat seeds were soaked in an acetone solution (250 mg/25 mL) of coumarin for two hours. After drying the

seed for several days in moving air, the seeds were set to germinate. Coumarin and other phenolic inhibitors were found to inhibit
germination only under the high moisture conditions, but in drier
sand had no effect or stimulated germination slightly. Acetone-
soaked seeds behaved essentially as the control, untreated seeds.

The phenomenon of water-sensitivity has been much studied in
barley. If semidormant barley is set to germinate in petri plates,
germination rate is maximal with 4 mL of added water; larger volumes
result in marked delay of germination (Briggs 1978; Pollock 1962).
Janson (cited by Briggs 1978) found that water sensitivity of barley
could be induced in the presence of added coumarin. Apparently, in
wheat, coumarin induces a condition that looks rather like water-
sensitivity in barley.

The pattern of response to coumarin might be consistent with
the needs of the developing grain. Where water is freely available,
as during grain development, germination of the embryo must be held
in check. At the same time, it would be advantageous that the inhibitor not act under natural germination conditions in the soil where
rate of water uptake is restricted. Barlow et al (1980) showed that
developing wheat grain maintained a constant ψ of about -10 bars
under conditions where the rachilla, glumes and flag leaf lost water
to the extent that ψ was lower than -30 bars. Grain ψ only declined
at maturity when grain was rapidly losing water.

Campbell (1958) found that in samples of Manitoba wheat, but
not of British wheats, there was very rapid hydration of the dorsal
endosperm during tempering of the grain. He reasoned that mechanical damage to the pericarp in the brush region could permit water to
move by capillarity beneath the pericarp. Following Campbell's procedure, when the brush end of individual kernels was immersed in a
dye solution, the dye entered under the pericarp of damaged grain
beginning in the brush region and moved rapidly along the dorsal
surface toward the embryo. Thus mechanical damage could be a source
of variability in germination studies. Seeds were placed, crease
down into petri dishes containing a 1:10 dilution of black ink
(Osmiroid). The ink does not appear to be toxic and the dyes do not
stain the tissues. Neepawa wheat showed dye-entry under the pericarp within five min. of placement in the dye. Seeds of Glenlea
wheat were hand picked from harvested heads. Under the same conditions no dye-entry was observed after 30 minutes. However, seeds
hand picked from heads of Neepawa wheat showed dye entry within a
few minutes. This suggested that while mechanical damage during
harvesting and handling may be involved, drying of the pericarp
under some conditions or in some cultivars could result in a crack
or tear in the pericarp in the brush region.

In many wheats the brush region is set off from the dorsal
region of the pericarp by the so-called "collar", which amounts to a
fold in the pericarp. According to Bradbury et al (1956) most
wheats have an air space in this region.

In most hexaploid wheats, the brush is a dense mass of very long hairs. Durum wheats have fewer and shorter hairs. The collar-brush region of seeds of durum wheat (cv Leeds) was examined under the scanning electron microscope. A crack or tear was observed in many of the seeds usually within the fold of the collar (Fig. 2). Similar breaks were seen in Neepawa seeds after the brush was burned away by passing the seed several times through a flame, but it is not certain that heating did not alter the structure.

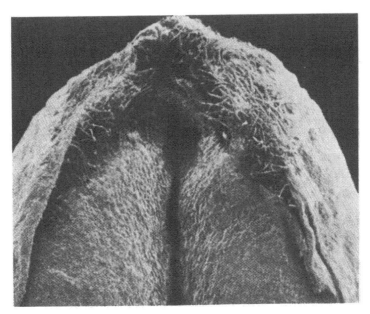

Fig. 2. Brush-collar area; crack on right edge.

The dye system was refined in the following ways. Petri plates were selected to be a few mm larger than the diameter of 9.0 cm (Whatman No. 1) filter discs. The paper expands slightly when wet and will buckle if the extra margin is not allowed. One to four filter discs were placed in the dishes which must be located on a horizontal surface. Four mL of dye solution was slowly pipetted into each dish and air bubbles were carefully worked out from beneath the paper. These operations assured a fairly constant depth of solution. The seeds were placed, crease down at least 1 cm apart and 1 cm away from the edge of the dish in order to prevent overlap of the various menisci. Finally, not more than 25 seeds were placed in each dish since water held in the meniscus around the seeds will lower the depth of the solution. Typical results are shown at 15 minutes after placement of the seeds (Table 1).

Table 1.   Number of Seeds (of ten) Showing Dye Entry Under The
            Pericarp in the Brush Region.

Filter Papers	Neepawa*	Leeds	Columbus	Neepawa**
1	3.7	2.3	2.0	3.0
2	3.7	2.0	0	0
3	0	0	0	0
4	0	0	0	0

* Column 1, bulk sample;
** Column 4, medium size.

The number of seeds showing dye-entry increased with incubation time but usually did not increase beyond 70% even after four hours. With three filter papers, dye entry was absent or very low. It was never observed with four papers in the dish. With 3 papers, the surface was barely wet.

These results are of a preliminary nature, but do show that dye can move beneath the pericarp in a significant number of seeds under usual seed germination conditions. Differences between varieties could be due to variation in shape or size of seed and in waxiness or roughness of the seed coat and brush. These variables could determine the extent and rate of invasion of the brush by the dye solution.

This pattern of water movement could be part of the explanation for the result obtained with coumarin. Unfortunately, with treated seed, acetone reduces the number of seeds showing dye entry but the coumarin (in acetone) treated seeds show a somewhat higher value. These effects are not understood, but it is possible that acetone-soaking causes relocation of surface waxes or phenols to the capillary space under the pericarp or acetone may cause a rearrangement of the microstructure.

In other work, it was found that seeds known to be dormant in the usual assay system, showed minimal dormancy in the sand system if less than 14 mL of water was present. This parallels the earlier results with coumarin. If water moved rapidly beneath the pericarp, it could transport inhibitors from the pericarp to the embryo. Accordingly, experiments were carried out at 20°C. Dormant and non-dormant Neepawa seeds were used and the sand contained 10 mL of water. Seeds were placed either on the sand surface or were inserted about half their length with the brush or the embryo end into the sand (Table 2).

Table 2.   Effect   of   Placing   Wheat   Kernels   on   Sand   With   Surface
Contact   or   Embryo   and   Brush   Ends   Inserted   in   the   Sand
Medium.

| | Hours to 50% Germination | | % Water at 50% Germination |
	Nondormant	Dormant	
Surface	56	55	54
Brush down	53	100	60
Embryo down	31	32	54

With the embryo down, germination is very rapid, possibly
because of the short path of water movement to the embryo. Dormancy
was only seen with the brush down seeds, suggesting that water moving along the pericarp was transporting inhibitor to the embryo.
Brush down seeds showed slightly greater water uptake which may represent water held beneath the dorsal pericarp. Similar results were
obtained in other varieties where dormant seed was available in sufficient quantity.

DISCUSSION

The important result in this paper would seem to be that dormancy, whether by added coumarin or natural post-harvest dormancy,
is most strongly expressed under conditions where the seed is in
contact with liquid water. We also find, as was indicated by MacKey
(1976), that natural dormancy is only expressed at temperatures
above 20°C. Since liquid water contacts and high temperature are
used in most assays of sprouting resistance, we wonder whether the
total range of dormancy has been exploited. Are there genes which
would condition dormancy at lower temperatures or on moist sand?

The discovery that dye-entry beneath the pericarp is apparently
involved in expression of dormancy and is associated with a cracking
of the collar structure, apparently during grain development, may
provide some insight into the problem of how rate and temperature of
grain drying can influence dormancy and germinability of intact
grain.

Bushuk and Winkler (1957) investigated adsorption of water
vapor to gluten, starch and flour. At moisture levels below 20%,
different polymers displayed different binding energies for water.
Plots of binding energy vs humidity were not smooth curves. Peaks
were located at different humidities in the different polymers.
This would suggest that during drying in the moisture region below
20% where conformational changes in the polymers would occur, relative rates of water loss from different polymers could change markedly with a fairly small change in temperature.

It is a fundamental of polymer science that polymers have a memory. Conformation and properties of a particular polymer sample will depend on its past history (Suggett 1975). If starch is suspended in water and heated, the granules will gelatinize at a fairly well defined critical temperature. On further heating, a sol will form with the individual molecules dispersed throughout the solvent. On rapid cooling, a fairly firm gel will form. On the other hand, with slow cooling, an insoluble, semi-crystalline precipitate is formed. In the gel, short regions of neighbouring chains are hydrogen bonded together, which stabilizes the structure. With slow cooling, enough time at higher temperature is allowed for the gradual alignment of polymer chains into semicrystalline arrays. Clearly, water retention and mechanical properties of a given weight of starch will be quite different in the two states. A similar result is obtained on neutralizing a solution of amylase in alkali at room temperature.

Tanaka (1981) has developed the full phase diagram for polyacrylamide-solvent interactions. Two points can be made from his paper. This system, like many other polymers, shows sharply defined critical temperatures for phase change. However, with gel pieces of 5 mm diameter, about the dimension of many seeds, temperature conditions sufficient to bring about the phase change would equilibrate throughout the system in minutes, but the phase change, whether swelling or collapse of the gel, can take many days to go to completion.

Campbell and Jones (1955) found that water movement to the cheek-center region of wheat endosperm during tempering was very slow, requiring 40 hrs for completion at 20°C. The rate increased about ten-fold between 20 and 43°C but higher temperature did not further increase the rate. They also found that treatment of the damped grain at 43°C resulted in a large increase in rate of water movement when the grain was returned to 20°C, but only if the time at high temperature exceeded 30 min. Thus, 43°C produced a phase change in some structure, not necessarily a membrane, which facilitated water movement.

A seed is a very complicated structure. It includes a number of distinct organs each of which will be structured differently both in respect to anatomical configuration and to chemical composition. Changes in moisture content of polymers will result in shrinkage or swelling. If these changes occur within a structural entity, stresses (of tension or of compression) will be generated, which can affect water relations locally, as well as in other tissues at some distance. Much of our current thinking about dormancy and sprouting resistance centers on membranes and inhibitor-promoter interactions, but there are a number of results in the literature which do not easily conform to these ideas. Water is required for germination, and for the biochemical changes involved. But it also has important interactions with structural components of the seed which "have a memory". The nature of this "memory" may well subtend germinability and dormancy.

Suggett (1975), in discussing the properties of polysaccharides in solution, concludes "Water - the original catalyst - intimately involved in all biological processes and yet, in the end, unchanged.".

REFERENCES

Barlow, E.W.R., Lee, J.W., Munns, R. and Smart, M.G. 1980. Water relations of the developing wheat grain. Aust. J. Plant Physiol. 7:519-525.

Belderok, B. 1976. Changes in the seed coat of wheat kernels during dormancy and after-ripening. Cereal Res. Commun. 4:165-170.

Bradbury, D., MacMasters, M.M. and Cull, I.M. 1956. Structure of the mature wheat kernel. Cereal Chem. 33:330-373.

Briggs, D.E. 1978. Barley. Chapman & Hall, New York.

Bushuk, W. and Winkler, C.A. 1957. Sorption of wheat flour starch and gluten. Cereal Chem. 34:73-86.

Campbell, J.D. and Jones, C.R. 1955. Effect of temperature on the rate of penetration of moisture within damped grain. Cereal Chem. 32:132-139.

Campbell, J.D. 1958. Effect of mechanical damage to wheat grains during scouring on their subsequent absorption of water during washing. Cereal Chem. 35:47-56.

King, R.W. and Gale, M.S. 1980. Preharvest assessment of potential α-amylase production. Cereal Res. Commun. 8:157-165.

MacKey, J. 1976. Seed dormancy in agriculture and nature. Cereal Res. Commun. 4:82-91.

Mitchell, B., Black, M. and Chapman, J.J. 1980. Drying and the onset of germinability in developing wheat grains. Cereal Res. Commun. 8:151-156.

Nicholls, P.B. 1979. Induction of sensitivity of gibberellic acid in developing wheat caryopsis: effect of rate of desiccation. Aust. J. Plant Physiol. 6:229-240.

Nicholls, P.B. 1980. Development of responsiveness to gibberellic acid in the aleurone layer of immature wheat and barley caryopses: effect of temperature. Aust. J. Plant Physiol. 7:645-653.

Pollock, J.R.A. 1962. The nature of the malting process. in A.A. Cook, Ed. Barley & Malt. Academic Press. 1962.

Stoy, V. and Sundin, K. (1976. Effects of growth regulating substances in Cereal seed germination. Cereal Res. Commun. 4:157-163.

Suggett, A. 1975. Polysaccharides. in F. Franks, Ed. Water, Vol. 4. Plenum Publishing Co., New York.

Tanaka, T. 1981. Gels. Scientific Amer. 244:124-138.

Wellington, P.S. and Durham, V.M. 1961. Studies on the germination of cereals. 3. The effect of the covering layers on uptake of water by the embryo of the wheat grain. Ann. Bot. 25:185-196.

Investigation of the Pre-Harvest Sprouting Damage Resistance Mechanisms in Some Australian White Wheats

D. J. Mares, *The University of Sidney,*
Plant Breeding Institute, Narrabri,
New South Wales, Australia

ABSTRACT

Northern N.S.W. hard wheats Kite, Shortim and Songlen, together with the advanced breeding line SUN 44E, have shown consistent differences in their susceptibility to pre-harvest sprouting damage over five years. Kite and SUN 44E are reasonably resistant, Shortim moderately resistant whilst Songlen is normally quite susceptible. In order to define the physiological and biochemical events which are characteristic of the more resistant cultivars, the process of sprouting in response to rain was divided into a number of steps which were then measured independently. Rates of movement of water into grains in intact ears, rates of germination of harvest mature grain and the rates of production of α-amylase all showed significant differences which in conjunction, could account for the observed varietal responses. Changes in germinability during ageing of the grain were responsible for the decline in resistance of all varieties after harvest maturity.

INTRODUCTION

Mature wheat crops in northern N.S.W. are commonly subjected to short periods of heavy rain, but in contrast to conditions which pertain to parts of Europe and North America, conditions conducive to sprouting do not normally persist for an extended time. High temperatures (30-40°C during the day and 15-25°C during the night) coupled with warm winds facilitate the rapid redrying of the crop. Consequently the levels of resistance required to protect the crops for the duration of the harvest are somewhat less than that considered necessary in some other parts of the world: a rather fortunate situation in view of Australia's tradition for cultivating white grained wheats.

Over the past 5 years it has been observed that

some commercial cultivars were consistently more resistant, although the mechanisms were not understood.   For
at least one of these varieties the resistance appeared
to approach the level required to protect crops under
most weather conditions likely to be experienced in the
northern Australian wheat belt.   Attention was focussed
on these varieties in view of our lack of practical
success with some reputed mechanisms of resistance [e.g.
bract inhibitors of germination from Kleiber described by
Derera and Bhatt (1980) and the gibberellic acid insensitivity of the extreme dwarf, Tordo, described by
Bhatt, Derera and McMaster, (1977)] and a general observation that, with a few exceptions, the tolerance to rain
of our breeding lines had been declining to alarmingly
low levels over the past few years.

The aim of this investigation was to study some
tolerant Australian wheats under dryland conditions
typical of the region and to define the mechanisms of
resistance or tolerance involved in order that they could
be more effectively retained in our spring wheat breeding
program.   The phenomenon of sprouting in the ear in response to rain was broken down into a series of interrelated steps (Figure 1) and the rates of some of these
steps estimated separately using experimental techniques
designed to minimize the effect of preceding events in
the sprouting sequence.

Figure 1

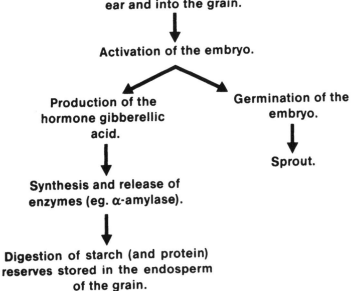

Figure 1. Schematic diagram of the sequence of events
which result in sprouting damage in wheat.

MATERIALS AND METHODS

Three commercial hard, white spring wheats; Kite (tolerant), Shortim (moderately resistant) and Songlen (susceptible); together with the advanced breeding line, SUN 44E, (tolerant) were grown at the Institute in 1981 under dryland cultivation.

Estimation of resistance to sprouting damage

150-200 heads of each variety were harvested at random from field plots (1 x 15 m, randomized complete block with four replications). At 0, 10 and 20 days after harvest maturity 30 heads were subjected to a standard weather treatment (50 mm of rain in 2 hours followed by 58 hours of high relative humidity at 20-25°C) in a controlled environment rain simulator (McMaster and Derera, 1975). The extent of weather damage was assessed by determining the falling number of weathered and non-weathered samples.

Estimation of the rate of movement of water into ears and grains

(i) Intact ears. Samples were removed at intervals from the rain simulator and the moisture contents of intact ears and grain, determined by drying at 50°C for 48 hours.
(ii) Isolated grains. Hand-threshed grains were incubated at 20°C in petri dishes on moistened filter paper and samples removed at intervals for moisture determination.

Estimation of germination rate

50 hand-threshed grains of each variety were surface sterilized in 1% Biogram for 10 minutes, washed three times with sterile deionized water, and incubated at 20°C in petri dishes with 4 ml sterile deionized water. The progressive percentage germination was recorded at daily intervals - germination being defined as the presence of a shoot and three distinct rootlets.

Estimation of the rate of α-amylase production

1.5 g of hand-threshed grain was surface sterilized with Biogram, washed, placed on filter paper in petri dishes together with 4 ml sterile deionized water containing Nystatin (100 μg ml^{-1}) and Streptomycin (100 μg ml^{-1}) and incubated at 4°C for 48 hours to abolish any dormancy. Plates were then transferred to 20°C and duplicate samples removed at intervals for estimation of α-amylase using the Phadebas method described by Barnes and Blakeney (1974).

Estimation of the amount of α-amylase required to reduce
the falling number to 250 seconds.

(i) α-amylase and falling number were determined on
samples taken at intervals from the rain simulator.
(ii) α-amylase was purified from germinated Kite grain by
the method described by Gibbons (1979) and aliquots of
known activity added to flours from non-weathered samples
just prior to the determination of falling number.

RESULTS

The results of a dryland trial from 1981 (Figure 2)
indicated that the resistance of SUN 44E was similar to
or slightly better than that of Kite despite having a
lower non-weather damaged falling number (390 sec for
SUN 44E compared with 490 sec for Kite, 440 sec for
Shortim and 450 sec for Songlen). Shortim was less
resistant whilst Songlen was the least resistant. This
pattern had been maintained over several years, however,
differences in absolute levels of resistance were noted
from trial to trial and year to year. SUN 44E was
distinguished by a lower rate of decrease in resistance
with time after maturity (Figure 2).
The initial movement of water into both ears and
hand-threshed grain was rapid with little significant
differences between varieties (Figure 3). By contrast
significant and consistent differences were observed in
the rate of movement of water into grains enclosed in
intact ears (Figure 3) - the rate being least for SUN 44E.
Of particular importance in terms of sprouting was the
time required to reach a water content of 45% dry weight;
the level required for germination to proceed (Mares unpublished data). For SUN 44E this period was almost
twice that for the least resistant variety Songlen.
Hand-threshed, mature grain of Shortim and Songlen
germinated more rapidly than SUN 44E which in turn germinated more rapidly than Kite (Figure 4). The germination rates for all varieties improved rapidly with
ageing. By 20 days after harvest the time required for

Figure 2. The response of four varieties of wheat 0, 10,
and 20 days post-harvest to a standard weather treatment.
Damage was estimated by the falling number method.
Figure 3. The rate of movement of water into intact ears
(I), isolated grains (II) and grains enclosed in intact
ears (III). Figure 4. The rate of germination of hand-
threshed wheat grain at harvest maturity at 20°C.
Figure 5. The rate of production of α-amylase by hand-
threshed wheat grain at 20°C following the removal of
dormancy by a 48 hour cold treatment. Explanation of
symbols used in Figures 2-6. (■) Songlen,
(▲) Shortim, (✽) Kite and (●) SUN 44E.

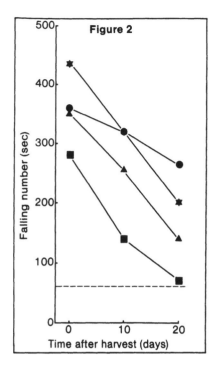

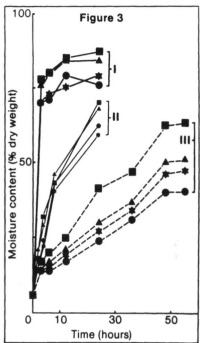

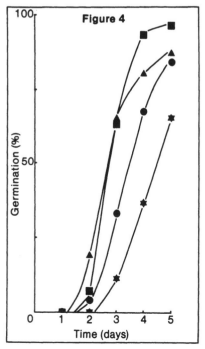

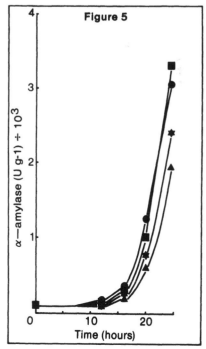

64

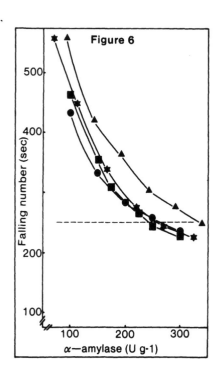

Figure 6. The amount of purified α-amylase required to reduce the falling number of non-weather damaged flour to 250 seconds.

50% germination of hand-threshed grains had decreased from 4.5 to 3.0 days in the case of Kite, 3.5 to 2.5 days for SUN 44E and from 2.7 to 2.0 days for both Shortim and Songlen.

Kite and Shortim showed significantly lower rates of α-amylase production than the other two lines (Figure 5). Slight differences were also noted in the amount of purified α-amylase required to reduce the falling number of sound flour, particularly of Shortim, to 250 sec. (Figure 6). Estimates obtained using purified α-amylase agreed very closely with figures obtained from a time course study in the rain simulation system where both α-amylase and falling number were determined (Mares, unpublished data).

DISCUSSION

Varietal differences were observed in the rates of all the components of the sprouting process that were measured experimentally, suggesting that resistance was a balance of several factors; the relative importance of each component changing with variety. Rates of other steps not measured here may also be very important. Songlen, the least resistant variety, was characterized by high rates for all component steps. The moderate resistance of Shortim appeared to depend on the slightly

lower rate of water uptake by the grain in the ear together with a reduced rate of α-amylase production and the
relative resistance of the "starch" to degradation in the
falling number apparatus. Kite, the most resistant commercial cultivar showed low rates for all the component
processes that were measured. The physiological and biochemical basis of the differences in the rates of water
movement into grains in intact ears remains unclear,
although, it appears to be an important component in the
resistance of SUN 44E. Water movement into both the non-
grain structures of the ear and isolated kernels, at
least in the initial few hours, was rapid and similar for
the different cultivars suggesting that varietal differences for grain enclosed in intact ears were determined
by the nature of the interface between the grain surface
and the interior surface of the glumes. The type of resistance shown by SUN 44E is quite attractive in view of
its simplicity and its relative insensitivity to changes
in the environment.

The rapid decline in resistance of all varieties
after maturity was closely related to changes in the
rates of germination of grain which accompanied post-
harvest after-ripening.

Having identified some mechanisms which appear to
confer tolerance, future efforts will be concentrated on
defining the effect of the environment on the different
components and investigating the practical application of
these types of resistance in our quality, spring wheat
breeding program.

REFERENCES

Barnes, W.C. and Blakeney, A.B., 1974. Determination of
    cereal alpha-amylase using a commercially available
    dye-labelled substrate Stärke 26:193-197.
Bhatt, G.M., Derera, N.F. and McMaster, G.J. 1977.
    Utilization of Tom Thumb source of pre-harvest
    sprouting tolerance in a wheat breeding programme.
    Euphytica 26:565-572.
Derera, N.F. and Bhatt, G.M. 1980. Germination inhibi-
    tion of the bracts in relation to pre-harvest sprout-
    ing tolerance in wheat. Cereal Res. Commun. 8:199-
    201.
Gibbons, G.C. 1979. Immunohistochemical determination
    of transport pathways of α-amylase in germinating
    barley seeds. Cereal Res. Commun. 8:87-96.
McMaster, G.J. and Derera, N.F. 1975. Methodology and
    sample preparation when screening for sprouting
    damage in cereals. Cereal Res. Commun. 4:251-254.

Effects of Various Saccharides on Gibberellic Acid Sensitivity of *Avena fatua* Seed

A.L.P. Cairns and O. T. de Villiers, *University of Stellenbosch, Republic of South Africa*

ABSTRACT

GA_3-mediated α-amylase synthesis in de-embryonated Avena fatua endosperm halves and in excised embryos with attached scutella, was influenced by the concentration of exogenous saccharides included in the incubation medium. The inclusion of low concentrations of mono- and disac= charides resulted in a stimulation of α-amylase synthe= sis, but higher concentrations of these sugars inhibited synthesis. The possibility of saccharides being involved in the control of α-amylase synthesis and dormancy is discussed.

INTRODUCTION

The genetic basis of GA_3 insensitivity has been well documented (Gale and Marshall 1975, Gale and Law 1976, and Flintham and Gale 1980) but very little work has been done to explain the physiological basis of this phenome= non. Evidence that the synthesis of gibberellin in ger= minating barley embryos was inhibited by a disturbance of the sugar metabolism was first presented by Radley (1969). She found that isolated scutella failed to produce gib= berellin if incubated on agar containing 2% sucrose or glucose although intact embryos did. Jones and Armstrong (1971) found that various osmotically active substances including glucose and sucrose, inhibited gibberellic acid - induced α-amylase synthesis in wheat aleurone layers. Nicholls (1979) showed that wheat caryopses from ears cul= tured on 0.4 or 4% sucrose, failed to produce α-amylase when incubated with GA_3. Cameron-Mills and Duffus (1980) demonstrated that when excised barley embryos were cul= tured on media containing 3% sucrose they germinated pre= cociously while if the sucrose level in the medium was increased to 12% the embryos failed to germinate com= pletely. Cairns (1982) found evidence to link gibberel= lic acid insensitivity with dormancy in Avena fatua. Caryopses from panicles which had been cultured on su=

crose solutions became progressively more dormant and more insensitive to gibberellic acid with increasing con=centration of sucrose in the culture solution. De-embryo=nated endosperm halves derived from whole plants, which were incubated with either 300 mM of sucrose or maltose, failed to respond to GA₃ but glucose and fructose at the same concentration did not induce gibberellic acid insen=sitivity. This study was consequently undertaken to exam=ine the effects of various saccharides on the insensitiv=ity of de-embryonated endosperm halves and excised whole embryos of Avena fatua.

MATERIALS AND METHODS

Experiments were conducted with a pure non-dormant line of Avena fatua derived from plants growing in infes=ted fields near Stellenbosch. Seed used in the experi=ments was grown in a glasshouse at 21/15°C day/night temp=erature and stored at 4°C. α-Amylase synthesis in de-embryonated endosperm half seeds harvested from whole plants grown in the glasshouse and incubated in the pres=ence of various saccharide solutions, was investigated under sterile conditions. De-embryonated endosperm half seeds were sterilized in 1% NaOCl (diluted "Jik") for 90 minutes. The seeds were then washed 10 times with sterile water, transferred to 50 ml Erlenmeyer flasks and incuba=ted for 48 hrs on a shakerbath at 30°C with 5 ml of the test solution containing 20 mM CaNO₃ and 10^{-8}M gibberellic acid. All solutions were sterilized by either autoclaving or by filtration through a 0.25 µm Millipore filter and all manipulations up to and including the incubation period were conducted in a laminar-flow sterile cabinet.

α-Amylase was assayed according to the method of Barnes and Blakeney (1974). Extracts were incubated with a highly specific dye-labelled substrate ("Phadabas" tab=lets, Pharmacia, Uppsala, Sweden) at 50°C for various periods depending on the α-amylase activity. The reaction was stopped by adding 1 ml 0.5 M NaOH. The samples then stood overnight to allow the undigested substrate to settle. A 5 ml aliquot of the supernatant was made up to 10 ml with sodium chloride/calcium acetate buffer and the absorbance determined at 620 nm. α-Amylase activity was expressed as enzyme units where one unit of amylase activ=ity is defined as the amount of enzyme catalyzing the hydrolysis of 1 µM glucosidic linkage per minute at 37°C.

For the embryo experiments seed was dehusked and imbibed in distilled water for two hours. The testa and pericarp covering the embryo and scutellum were stripped off under a microscope using a scalpel and tweezers. The scutellum was then peeled away from the endosperm. All traces of endosperm tissue were removed from the scutel=la. The synthesis of α-amylase by whole (axis + scutel=

68

lum) embryos was investigated by incubating them on an agarose/blue starch substrate. The α-amylase specific dye-labelled substrate was a gift from Pharmacia, Uppsala Sweden. The substrate was formulated as described by Hejgaard and Gibbons (1979) but was supplemented by gib= berellic acid at 10^{-6}M and varying ammounts of sucrose. The substrate was cast on to 8 x 5 cm glass plates. Af= ter the plates had cooled, the whole embryos were placed on the surface and covered with a second glass plate which was lightly pressed down to exclude air. The sub= strate containing the embryos was thus sandwiched between the two plates. The plates were incubated in a humid at= mosphere at 15°C for 120 hrs. The reaction was stopped by immersing the plates in 5% acetic acid for two hours. α-Amylase activity was determined by measuring the dia= meter of the zone of digested substrate accurately with a slide calliper.

RESULTS

α-Amylase synthesis by de-embryonated endosperm halves incubated with gibberellin and the various saccha= rides is illustrated in Fig. 1. The two monosaccharides,

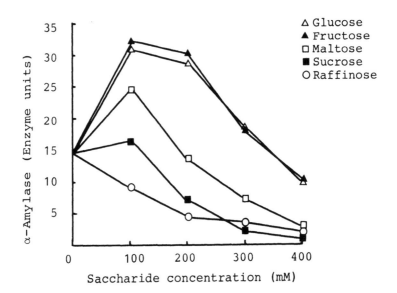

FIG. 1   Effects of saccharide concentration on
          α-amylase synthesis in de-embryonated
          endosperm halves of Avena fatua

glucose and fructose both stimulated α-amylase production
appreciably at 100 mM. The monosaccharides also stimula=
ted α-amylase synthesis at the 200 and 300 mM levels but
resulted in a significant inhibition at 400 mM. Maltose
was stimulatory to α-amylase synthesis at 100 mM but was
inhibitory at the higher concentrations. Sucrose showed
only slight stimulation at 100 mM, but was inhibitory at
the higher concentrations. Raffinose inhibited synthesis
at all concentrations. In further experiments (results
not shown) it was found that the optimum concentration of
sucrose for gibberellin stimulated α-amylase synthesis
was 75 mM. The order of stimulation achieved at this con=
centration was similar to that obtained with glucose and
fructose at 100 mM.

The effect of various concentrations of sucrose in=
cluded in the agarose/blue-starch substrate on α-amylase
synthesis of whole embryos is presented in Fig. 2. Opti=
mum stimulation was noted at 75 mM. Stimulation was also
noted at 150 mM but 300 and 600 mM were strongly inhi=
bitory.

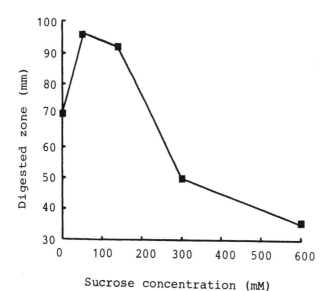

FIG. 2   Effects of sucrose concentration on
α-amylase synthesis in excised whole
embryos of Avena fatua

DISCUSSION

Clearly all the saccharides examined in this study were able to influence the sensitivity of the aleurone tissue to gibberellic acid. The fact that all the sac=charides, with the exception of raffinose, stimulated α-amylase synthesis at the lower concentrations would seem to indicate that they are operating as physiological=ly active substances rather than as inhibiting osmotica as suggested by Jones and Armstrong (1971). It is also interesting to note that it is α-amylase synthesis and not activity that is influenced by saccharide concentra=tion. At the higher concentrations the saccharides were probably triggering a negative feedback type of reaction. The large difference in activity between the two monosac=charides and the di- and trisaccharides cannot be ex=plained at this stage and further experiments will have to be conducted to resolve the matter.

The effect of sucrose on the potential of whole em=bryos to synthesise α-amylase was very similar to that on the aleurone layer. In both cases the optimum sucrose concentration was 75 mM. The synergistic effect of su=crose on the ability of low concentrations of GA₃ to sti=mulate germination of dormant <u>Avena fatua</u> embryos was commented on by Naylor and Simpson (1961). They found that embryos which had been stored for 18 months showed an optimum sucrose concentration for germination of 2%. Higher concentrations were inhibitory. However, the op=timum sucrose concentration for seeds which had been stored for only 5 months was 10%. It therefore seems likely that stage of after-ripening will affect the re=sponse of the embryo to sucrose. Although there appears at this stage to be no direct relationship between GA₃ insensitivity and dormancy the fact that sucrose can in=fluence both is interesting. The synthesis of germina=tive α-amylase is assumed to be a post germinative phe=nomenon but Gibbons (1981) has shown that in barley, the synthesis of α-amylase by the scutellum is an important event in the early germinative stages. Although no di=rect evidence can be presented it seems plausible that the variation of saccharide concentration available to the embryo and surrounding tissues could form a link in a system which controls both α-amylase production and dormancy either directly or indirectly.

ACKNOWLEDGEMENT

This research is part of the senior author's Ph.D.-thesis at the University of Stellenbosch.

REFERENCES

Barnes, W.C. and Blakeney, A.B. 1974. Determination of cereal alpha-amylase using a commercially available dye-labelled substrate. Die Stärk 26: 193-197.

Cairns, A.L.P. 1982. Induction of dormancy and gibber= ellic acid insensitivity in Avena fatua. Proceed= ings of the Fourth National Weeds Conference of South Africa, 75-81.

Cameron-Mills, V. and Duffus, C.M. 1980. The influence of nutrition on embryo development and germination. Cereal Res. Commun. 82:143-150.

Flintham, J.E. and Gale, M.D. 1980. The use of Gai/ Rht3 as a genetic base for low α-amylase wheats. Cereal Res. Commun. 82:283-290.

Gale, M.D. and Law, C.N. 1976. Proceedings of 'Sympos= ium on Genetic Diversity in Plants', 133-151.

Gale, M.D. and Marshall, G.A. 1975. The nature and genetic control of gibberellin insensitivity in dwarf wheat grain. Heredity 35:55-65.

Gibbons, G.C. 1981. On the relative role of the scu= tellum and aleurone in the production of hydrola= ses during germination of barley. Carlsberg Res. Commun. 46:215-225.

Hejgaard, J. and Gibbons, G.C. 1979. Screening for α-amylase in cereals. Improved gel-diffusion assay using a dye-labelled starch substrate. Carlsberg Res. Commun. 44:21-25.

Jones, R.L. and Armstrong, J.E. 1971. Evidence of osmotic regulation of hydrolytic enzyme production in germinating barley seeds. Plant Physiol. 48:137-142.

Naylor, J.M. and Simpson, G.M. 1961. Dormancy studies in seeds of Avena fatua 2. A gibberellin sensitive inhibitory mechanism in the embryo. Can. J. Bot. 39:281-295.

Nicholls, P.B. 1979. Induction of sensitivity of gibberellic acid in developing wheat caryopses: Effect of rate of desiccation. Aust. J. Plant Physiol. 6:229-240.

Radley, M. 1967. Site of gibberellin-like substances in germinating barley embryos. Planta 75:164-171.

Effects of Fertility and Rain Simulation During Grain Fill on Protein Content, Starch Quality, and Alpha-Amylase Activity in Winter Wheat

A. J. Ciha[1] and W. A. Goldstein[2],
*USDA-ARS and Washington State
University, Pullman, Washington, U.S.A.*

SUMMARY

Two winter wheat cultivars('Moro'-a soft white club and 'Mironovskaya 808'-a hard red common) were grown using various nitrogen fertilization levels and with artifical wetting of the heads at various stages in grain development. Changes in protein content and alpha-amylase activity were determined. The wetting treatments significantly influenced grain protein, alpha-amylase activity, and amylograph values; however, the effects of wetting treatments were cultivar dependent while the fertilizer treatments only influenced grain protein and was also cultivar dependent. Head wetting resulted in an increase in alpha-amylase activity and a decrease in amylograph values for Moro but not Mironovskaya 808.

INTRODUCTION

Pre-harvest sprouting can be a serious problem in wheat producing areas where rain precedes harvest. During a normal growing season in the Pacific Northwest, rainfall during June, July, and August is normally very limited, thus sound quality wheat can usually be harvested. On the other hand a week or more of rainfall may occur during the grain filling period. Also, the use of overhead sprinklers for irrigation of wheat in the drier regions of the Pacific Northwest has increased. Short periods of rainfall may have little influence on grain alpha-amylase if air moisture levels are not raised for a long period of time (Olered, 1967). Questions as to the changes in starch quality and alpha-amylase levels due to natural or artifical rainfall during various stages of grain development have led to this experiment.

Contribution from USDA/ARS in cooperation with the College of Agriculture Research Center, Washington State University, Pullman, WA 99164. Scientific Paper No. SP 6202.

[1]Research agronomist, USDA-ARS and Washington State University

[2]Graduate student, Washington State University

Producers have also increased the quantity of nitrogen fertilizer applied to wheat to increase grain yields; however, little is known about what changes in pre-harvest sprouting and starch quality occur as nitrogen fertility is increased.

This experiment was established to examine the effects of artifical wetting of wheat heads at various stages of seed development and of various nitrogen levels on alpha-amylase and starch quality of winter wheat.

MATERIALS AND METHODS

A field experiment was established on the Spillman Agronomy Farm near Pullman, Washington in the fall of 1980 on a Palouse silt loam(Pachic Ultic Haploxerolls) to examine the influence of fertility levels and rain simulation during grain fill on starch quality of two winter wheat cultivars. The winter wheat cultivars, 'Moro'(a soft white club) and 'Mironovskaya 808'(a hard red common), were seeded on 4 November 1980 at a rate of 67 kg/ha in rows 6 m. long and 30 cm. wide. Standard production practices for the area were used in establishing the experiment.

The experiment was a split-split plot, randomized complete block design, consisting of three replications, with five watering treatments as the main plot, five fertilization treatments as subplots, and two winter wheat cultivars as sub sub plots.

The five watering treatments consisted of a control with no watering, watering during either the milky, soft dough, or hard dough stage, and watering during all three stages of grain development. Each watering plot was 15 x 6 m. in size and sprinklers, which produced a square shaped watering pattern, were at a height of 1.5 m above the soil surface. Watering was applied for 30 minutes in the evening at 2000 hour and the heads usually remained wet until morning. The rate of flow from the sprinklers was approximately 7 liters of water/m^2 in the 30 minute period.

Within each watering treatment there were five fertility treatments consisting of a control with no fertilizer, a 90 or 180 kg/ha of nitrogen as NH_4NO_3(N), and a 18 or 36 mt/ha of dairy manure(FYM). The manure was about six months old and well decomposed. Analysis of the manure showed 0.65+.13% nitrogen for the moist material which resulted in the 18 and 36 mt/ha of manure having 117 and 234 kg of nitrogen/ha, respectively. Manure and fertilizer treatments were spread by hand and disced into the soil.

Samples of three hundred seeds were taken from each of the wheat yield samples and inspected for incipient sprouting. A seed was considered sprouted if the seed coat was ruptured in the area around the embyo or if there were any visible signs of coleoptile or seminal root emergence.

Seed germination tests were carried out directly after harvest on hand threshed samples of 10 randomly selected heads per treatment. Fifty seeds were placed on a 8.3 cm diameter piece of seed germination blotting paper and two 8.3 cm diameter pieces of Kimpak paper inside a 9 cm petri dish, watered with 10 ml of distilled water, and germinated at 15 and 30 C. The number of germinated seeds were counted from day 3 through day 7 to determine a promptness index(George, 1967) and percent germination. Germination was defined as emergence of 2 mm of a seminal root.

Seed protein percentage was determined on a subsample from the grain yield sample ground through a UDY Cyclone Mill containing a 0.5 mm screen. Kjeldahl protein content was determined by AACC approved methods.

Wheat flour samples were tested for starch viscosity using the standard A.A.C.C. amylographic method 22-10, which ultilizes a ratio of 65 gm of flour to 460 ml of water. Flour for amylographic work was obtained by milling 150 gm of seed on an Quadrumat IV research mill. Alpha amylase levels were determined by using A.A.C.C. method 22-06, which is a photometric method measuring dye release from Cibacron-blue amylase substrate tablets using 200 mg of flour on an alpha-amylase Analyzer(Model L3, D & S Instruments, Ltd., Pullman, WA.) with a reaction time of three minutes at 50 C. From a standard curve of malted barley the alpha amylase concentration was quantified in dextrinizing units per gram of dry weight of flour.

RESULTS AND DISCUSSION

Examination of the seed at harvest showed that Moro had a low level of visual sprouting while Mironvskaya 808 had no visual sprouting(Table 1). While the watering and fertility treatments were not significantly different, there was a small increase in the quantity of visual sprouting with Moro at the soft dough and milky-hard dough wettings.

Table 1. Percent visual sprouting at harvest for two winter wheat cultivars after artifical wetting of the heads at various stages of seed development.

Cultivar	Control	Milky	Soft Dough	Hard Dough	Milky-Hard Dough
			----%----		
Moro	0.1	0.4	1.0	0.3	1.3
Mironovskaya 808	0	0	0	0	0

There were no significant watering or fertility
effects on percent germination and germination index
for seeds germinated at 15 or 30 C; however, the
cultivars were significantly different with respect to
their level of seed dormancy(Table 2). Both the
germination index and percent germination indicated
that Moro was the most susceptible and Mironovskaya
808 the least susceptible to pre-harvest sprouting.

Table 2. Percent germination and promptness index
for winter wheat cultivars germinated at
15 and 30 C.

Cultivar	Percent Germination		Germination Index	
	15C	30C	15C	30C
Mironovskaya 808	93b*	6b	105b	5b
Moro	100a	77a	188a	112a

*5% Level of significance according to Duncan's
Multiple Range Test within a column.

Examination of the cultivars indicated that
Mironovskaya 808 was not generally influenced by the
head wetting or fertilizer treatments while Moro was
significantly influenced by both (Table 3 and 4).
Mironovskaya 808 had a higher grain protein content, a
higher amylograph value, and a lower alpha-amylase
activity than Moro. Increasing fertilizer and manure
applications tended to raise amylograph levels for
Mironovskaya 808. However, the difference was only
significant between the control and 36 mt/ha FYM
treatments. The continuous wetting with Moro resulted
in a significant decrease in the amylograph values and
a higher alpha-amylase activity than the control; the
other watering treatments gave parameter values in
between these two. These findings indicate that even a
small quantity of head wetting can substantially
change alpha-amylase activity and amylograph values
for some cultivars. Wetting the heads at the early
stages of seed development generally resulted in the
greatest increase in grain protein. Increasing the
fertilizer level significantly increased the grain
protein with Moro but not with Mironovskaya 808. The
grain protein was increased more by the application
of N than FYM.

Table 3. Effects of fertilizer treatments on grain protein, alpha-amylase activity, and amylograph values for two winter wheat cultivars.

Cultivar	Fertilizer treatment	Grain protein	Alpha-amylase	Amylograph
		%	Du/gdw	B.U.
Moro	Control	10.9[+]	.72	283
	90 kg/ha N	11.8	.72	261
	180 kg/ha N	12.2	.88	252
	18 mt/ha FYM	11.2	.91	227
	36 mt/ha FYM	11.4	.90	227
	mean	11.5a	.82a	250b
Mironovskaya 808	Control	12.0	.59	767
	90 kg/ha N	11.9	.58	776
	180 kg/ha N	12.2	.61	845
	18 mt/ha FYM	12.3	.59	806
	36 mt/ha FYM	12.4	.64	865
	mean	12.2a	.60b	812a
L.S.D.(0.05%) between fertility treatments within a column		0.5	.26	91

+ 5% Level of significance according to Duncan's Multiple Range Test for cultivar means.

Amylograph values were negatively correlated (r= -.47) (p < 0.01) with alpha-amylase activity, while protein content was positively correlated (r=.31) (p < 0.01) with amylograph values but not alpha-amylase activity (r= -.01). Bhatt et al. (1980) reported no significant effect of nitrogen fertilizer on sprouting. However, they did have a cultivar x nitrogen interaction for Falling Numbers but not for alpha-amylase activity; for some cultivars nitrogen fertilization decreased the Falling Number value. Moss (1967) reported that protein content was negatively correlated with alpha-amylase activity. However, Porsche and Schmieder (1980) reported increased protein content and alpha-amylase activity as nitrogen fertilization increased. Huang and Varriano-Marston (1980) found a low negative correlation between grain protein and alpha-amylase activity (r=-.49) and they suggested that increased protein synthesis in the grain may not parallel increased enzyme synthesis. Increased use of over-head irrigation systems may influence the enzymatic activity of the grain; however, this change will be cultivar dependent.

Table 4. Effects of head wetting at various stages of seed development on grain protein, alpha-amylase activity and amylograph values of two winter wheat cultivars.

Cultivar	Wetting treatment	Grain protein	Alpha-amylase	Amylo-graph
		%	D.U./gdw	B.U.
Moro	Control	11.3+	.61	364
	Milky	11.8	.82	261
	Soft-dough	11.5	.70	230
	Hard-dough	11.2	.94	233
	Milky-hard dough	11.6	1.06	160
	mean	11.5b	.82a	250b
Mironovskaya 808	Control	12.0	.55	845
	Milky	12.4	.66	780
	Soft-dough	12.3	.62	817
	Hard-dough	12.0	.61	835
	Milky-hard dough	12.1	.59	782
	mean	12.2a	.60b	812a
L.S.D.(0.05) between watering treatments within a cultivar		0.5	.26	91

+ 5% Level of significance according to Duncan's Multiple Range Test for cultivar means.

Also, the occurence of the low amylograph values for Moro even though the alpha-amylase activity was relatively low suggest the possibility of erroneously classifying sound wheat as damaged based on amylograph values. The low amylograph value obtained for Moro with wetting from the milky through the hard dough stage (160) might have precluded the acceptance of the material for baking purposes. Our findings are in agreement with Moss (1967) who suggests the need to confirm amylograph values with enzymatic activity.

REFERENCES

American Association of Cereal Chemists. 1962. Approved Methods of the A.A.C.C. The Association: St. Paul, MN.

Bhatt, G.M., G.M. Paulsen, K. Kulp and E.G. Heyne. 1981. Pre-harvest sprouting in hard winter wheats: Assessment of methods to detect genotypic and nitrogen effects and interactions. Cereal Chem. 58:300-302.

George, D.W. 1967. High temperature seed dormancy in wheat. Crop Sci. 7:249-252.

Huang, G. and E. Varriano-Manrston. 1980. a-Amylase activity and pre-harvest sprouting damage in Kansas hard white wheat. J. Agric. Food Chem. 28:509-512.

Moss, H.J. 1967. Flour paste viscosities of some Australian wheats. J. Sci. Fd. Agric. 18:610-612.

Olered, R. 1967. Development of a-amylase and falling number in wheat and rye during ripening. Publication of the Dept. of Plant Husbandry at the Agric. Univer. of Sweden. Uppsala. No. 23.

Porsche, W. and W. Schmieder. 1980. Problems and results of breeding winter wheat with high resistance to sprouting. Archk. Zuchtungsforsch., Berlin 10:295-300.

Falling Numbers and Alpha-Amylase in Sawfly-Resistant Wheats

T. N. McCaig and R. M. De Pauw,
Research Station, Agriculture Canada,
Swift Current, Saskatchewan, Canada

SUMMARY

Two experiments were carried out to determine the ancestral source of the sprouting susceptibility problem experienced by the sawfly resistant wheats Chester and Canuck. The first was a field experiment which included seven wheats from the sawfly resistant breeding program while the second was carried out in a growth chamber and included twelve wheats from the same program. Although the results are not conclusive, they suggest that the sprouting susceptibility may have been inherited from S615, the original source of the stem solidness in this program. Although there are many intervening crosses between S615 and Chester/Canuck, the main selection criterion has been for stem solidness which may have retained many characteristics associated with S615. Evidence from the growth chamber experiment also indicates that the low falling numbers of S615 may be partially due to starch characteristics.

INTRODUCTION

For the past forty years in Canada, solid-stem wheats have been developed in order to combat the wheat-stem sawfly (Cephus cinctus Nort.). The original source of stem solidness, for this program, was the line S615. S615 was subsequently crossed with bread wheats, such as Thatcher, to incorporate better quality and yield characteristics in solid-stem cultivars.

In the past few years more emphasis has been placed upon ensuring good falling numbers (FN), and, therefore, low alpha-amylase levels, in Canadian wheat for export. However, the two most popular sawfly resistant wheats grown in Saskatchewan, Chester and Canuck, both tend to have high alpha-amylase activity when subjected to wet harvest conditions.

The present study was undertaken with the hope of identifying the ancestral source of the high alpha-amylase problems of Chester and Canuck wheats.

MATERIALS AND METHODS

In 1980, a field experiment was carried out at Swift Current, Saskatchewan which included seven red and two white wheats (Triticum aestivum L.). The reds included the solid-stem wheats Canuck, Chester, S615, and Leader and the hollow-stem wheats Thatcher, Apex, and Mida/Cadet. The white wheats included were NB 112 and 7722. Four replicates of each wheat were grown in a randomized complete block. Plots were four rows with 0.23 m spacing and 3 m long.

As each wheat matured ($\leq$20% moisture, judged by loss of chlorophyll and seed hardness) head samples were cut leaving 8 to 10 cm of stem. These heads were then stored at constant 20°C. At grain maturity, and at weekly intervals for six subsequent weeks, twenty heads /rep were placed in the rain simulator for two days. The rain simulator was of the type described by McMaster and Derera (1976). Heads in the rain simulator received 50 to 60 mm of water over a three-hour period. The humidity was maintained at 100%, and temperature at 18°C throughout the two days. At weeks seven and eight, the heads were left in the rain simulator for three days.

In a second experiment, fifteen red wheats and two white wheats were grown in pots in a growth chamber maintained at 21°C (day, 16 hr light)/18°C (night). Plants were grown in a randomized complete block. At daily intervals, as heads matured, half of the mature heads were removed and stored at -20°C while the remaining half were tagged and left for an additional three weeks before being stored at -20°C. Heads were later placed in the rain simulator and maintained as described earlier for sixty hours.

Alpha-amylase was measured using the gel-diffusion technique described by Hejgaard and Gibbons (1979).

RESULTS AND DISCUSSION

Since the sprouting susceptibility problem is common to both Chester and Canuck wheats, the 1980 field experiment included the ancestral parents common to both wheats (i.e., S615, Thatcher, Apex, Mida/Cadet). When these wheats were subjected to two days in the rain simulator at grain maturity, and at six weekly intervals thereafter, the dormancy changes (as judged by FN) were relatively linear with time. For ease of comparison, therefore, linear regression lines were calculated from the data of the first six weeks (Fig. 1). At seven and eight weeks after maturity heads were left in the rain simulator for three days. However, this period was found to eliminate most of the varietal differences.

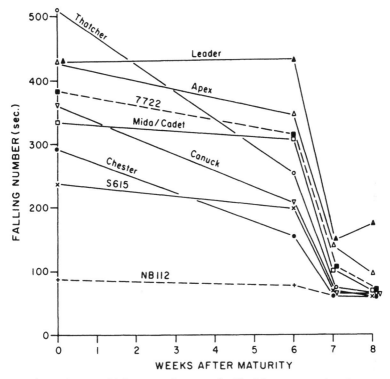

Fig. 1. Falling numbers of field grown wheats
following two days (wk 0-6) or three days
(wk 7-8) in a rain simulator

As expected, Chester and Canuck exhibited relatively low FN throughout, although Canuck maintained higher values than Chester. Of the common parents, Thatcher, Apex, and Mida/Cadet tended to have higher FN than Canuck and Chester. However, S615 had FN as low as, or lower, than Chester.

The rates at which the dormancy decreased over the six weeks were also different for the various wheats. Thatcher, which had the highest FN at maturity, displayed the most rapid decline. S615, Mida/Cadet and Leader were fairly stable over this period, while Apex, Canuck and Chester demonstrated intermediate rates of decline.

Leader is a recently licensed solid-stem wheat which incorporates the good sprouting resistance of

Chris (De Pauw et al. 1982). The high and stable dormancy of Leader should dramatically improve the reputation of Canadian solid-stem wheats.

Also included in this study, for comparison, were two white wheat lines. Two days in the rain simulator were sufficient to decrease the FN of NB 112 to less than 100. In fact, sprouting was visible for this line. However, the other white line, 7722, exhibited better FN than the red wheats S615, Chester, Canuck and Mida/Cadet. This wheat is presently being used in the Swift Current breeding program aimed at developing a sprouting resistant white wheat.

In the growth chamber experiment, heads were placed in the rain simulator both at maturity and at three weeks after maturity. This experiment also included a few more of the intervening solid-stem parents than were grown in the field. The falling numbers for these wheats (Fig. 2 and 3) represent averages for three separate treatments in the rain simulator.

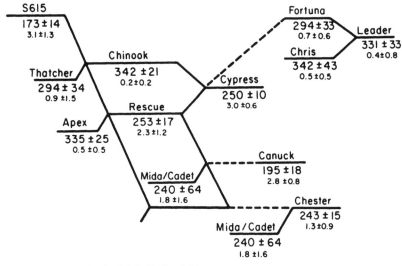

(NB112 = 221 ±38 ; 7722 = 346 ±20)
     2.7±1.4   0.6±1.0

Fig. 2. Falling numbers and alpha-amylase of wheats at maturity following rain simulator treatment (from growth chamber expt.). Numbers directly below varieties are falling numbers ±S.D; smaller numbers below represent alpha-amylase activity

When subjected to rain at maturity, the wheats recorded higher FN than when the procedure was repeated three weeks later, indicating that all of the wheats had some dormancy. At maturity, S615 and Canuck recorded

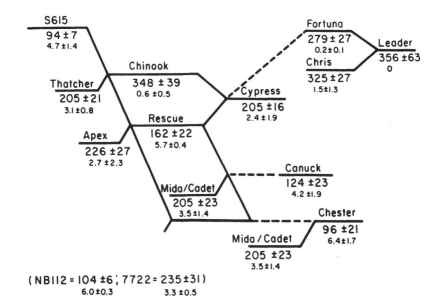

(NB112 = 104 ±6 ; 7722 = 235±31)
          6.0±0.3   3.3 ±0.5

Fig. 3. Falling numbers and alpha-amylase of wheats
at maturity + three weeks (details same as Fig. 2)

the lowest FN (<200); Chinook, Rescue, Mida/Cadet,
Cypress and Chester comprised an intermediate group
(~250); Thatcher, Apex, Chinook, Fortuna, Chris and
Leader yielded the highest values (≳300).
    Three weeks after maturity S615 and Chester both
had FN below 100, while Canuck was only slightly higher
(124), followed by Rescue (162). The intermediate group
(~200) included Thatcher, Apex, Mida/Cadet and Cypress
while Fortuna, Chris, Leader and Chinook continued to
show good dormancy (≳300).
    The white lines, NB 112 and 7722, were also
included in this experiment for comparison. At maturity
7722 retained a FN as high as any of the red wheats
tested and even NB 112 maintained a higher FN than the
red wheats Canuck and S615. Three weeks later 7722 continued to register reasonable FN while NB 112 had lost
dormancy.
    An indication of the alpha-amylase level of the
wheats from the growth chamber experiment was obtained
using the agar-plate method (Hejgaard and Gibbons,
1979). Athough there was considerable variability among
treatments, the measurements roughly followed the FN
(Fig. 2 and 3). However, when comparing wheats there
appeared to be some discrepancies between FN and alpha-
amylase. For example, at maturity+3 weeks, Rescue maintained higher falling numbers than S615 while also having higher alpha-amylase levels.

The relationship between FN and alpha-amylase activity for all wheats from the growth chamber experiment is shown in Fig. 4. The solid regression line was calculated using data from all wheats but excluding values with no measurable alpha-amylase activity (i.e., well dia. = 0). The second regression line (broken line) was calculated using only values from S615. The S615 line deviates considerably from the overall regression line. Although the results are somewhat preliminary, this suggests that for a given level of grain alpha-amylase, S615 has a much lower FN than the other wheats included. Such a relationship has been reported previously for different wheat classes (Mathewson and Pomeranz 1978). Most of the values for Canuck and Chester fall between the two lines which further suggests that the low FN of these cultivars may be partially due to a combination of starch characteristics and alpha-amylase activity.

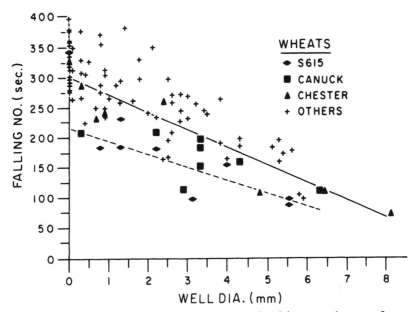

Fig. 4. Relationship between falling number and alpha-amylase activity (well dia.) for sawfly wheats. The solid regression line includes data from all wheats; the broken line includes data for S615 only

From these experiments it is tempting to speculate that the low FN of Chester and Canuck, resulting from adverse weather, may be partially attributable to S615, the original source of stem solidness in this program.

Although extensive crossing separates these wheats from S615, the major selection criterion has always been for stem solidness. This means that it is possible that many genes from S615 may have been retained throughout. Genes controlling sprouting susceptibility may be linked with some of these genes or otherwise inadvertently selected.

REFERENCES

De Pauw, R.M., McBean, D.S., Buzinski, S.R., Townley-Smith, T.F., Clarke, J.M. and McCaig, T.N. 1982. Leader hard red spring wheat. Can. J. Plant Sci. 62:231-232.

Hejgaard, J. and Gibbons, G.C. 1979. Screening for alpha-amylase in cereals. Improved gel-diffusion assay using a dye-labelled starch substrate. Carlsberg Res. Commun. 44.

Mathewson, P.R. and Pomeranz, Y. 1978. On the relationship between alpha-amylase and falling number in wheat. J. Food Sci. 43:652-653.

McMaster, G.J. and Derera, N.F. 1976. Methodology and sample preparation when screening for sprouting damage in cereals. Cereal Res. Commun. 4:251-254.

Section II
Chemistry of Pre-Harvest Sprouting

Recent Progress in the Biochemistry of Immature Cereal Grains in Relation to Pre-Harvest Sprouting

C. M. Duffus, *School of Agriculture,*
University of Edinburgh, Edinburgh, Scotland

SUMMARY

Some of the events of grain development in cereals which may be concerned in the control of growth, maturation and germination of the embryo are described. These include the biochemical and morphological changes occurring in the embryo, aleurone and outer layers of the developing caryopsis. The possible roles of sucrose (osmotic pressure), abscisic acid (ABA) and gibberellic acid (GA) are discussed.

INTRODUCTION

This review aims to describe some recent advances made in our understanding of the events occurring in cereal caryopses which may relate to pre-harvest sprouting. Since sprouting takes place before full grain maturity it is clear that the fundamental processes involved relate to those concerned in grain growth and differentiation and may be distinct from those concerned in germination. Much of the work relating to pre-harvest sprouting in cereals has been concerned with the appearance of increased α-amylase activity in immature caryopses. This is largely because this enzyme is characteristically associated with starch degradation in the endosperm following germination. However the release of amylolytic activity occurs relatively late in germination and it is likely that the key events involved in the onset of premature sprouting are associated with the embryo itself. Consequently in the discussion which follows some emphasis will be put on those changes accompanying grain maturation which may be concerned in the control of embryogenesis.

EMBRYO EVELOPMENT

After fertilisation the endosperm nucleus undergoes free nuclear division. Cell wall formation follows and in barley, maize, wheat and rye is generally complete within around seven days. The first divisions of the zygote take place after at least four nuclear divisions in the endosperm and initially give rise to the proembryo. Following further cell division the embryo is formed. This is located in the dorsal side at the base of the caryopsis and is effectively embedded in the endosperm. It becomes differentiated into a scutellum, which lies against crushed endosperm cells, and a radicle and plumule which are separated from the nucellar epidermis by a layer, one cell thick, of modified aleurone (see Duffus and Cochrane, 1982). As development proceeds the embryo grows steadily

89

within the endosperm. In barley it reaches a maximum size at about 45 days after anthesis but dry weight increases throughout development, reaching a maximum value in barley only by maturity (Ahluwalia, 1980). Thus although the rate of increase in number and/or size of embryo cells may have largely ceased by two thirds of the way through grain development, the cells continue to increase in density as storage material accumulates. By 21 days after anthesis, cells in the root region of developing barley embryos are thin-walled with large vacuoles (Ahluwalia, 1980). The cytoplasm is dense with numerous ribosomes and mitochondria with few cristae. There are also lipid bodies and plastids some of which contain starch granules. By 37 days after anthesis the cells are larger, with thicker walls and there is considerable accumulation of storage products within lipid bodies, amyloplasts and protein bodies. The mitochondria are widespread and numerous and contain many cristae but are rather smaller than those in younger embryos. Excised isolated barley embryos respire throughout development (Ahluwalia, personal communication) indicating that these mitochondria are metabolically active at least *in vitro*. Whether this is the case *in vivo* depends on the availability of oxygen. Supplies of oxygen to the developing embryo may be adequate early in development but may be limited at later stages when the cross cells of the pericarp no longer produce oxygen in photosynthesis (Duffus and Cochrane, 1982). Nutrients apparently enter the embryo by an apoplastic route since no plasmodesmata have been observed in the outer walls of the zygote nor in the outer epidermal walls of the embryo (Norstog, 1972; Ahluwalia, 1980). It is not known whether all nutrients and water reach the embryo via the endosperm or whether there is an additional independent supply from the rachilla along the base of the pericarp. Thus the concept that the embryo may be dependent on the endosperm for nutrients during development is based principally on guesswork. It is not necessarily the case therefore, that conditions in the endosperm affect embryo growth and development.

CHANGES IN THE MATERNAL TISSUES

Endosperm development is associated with changes in the surrounding maternal tissues (Duffus and Cochrane, 1982). In barley, oats, rye and wheat, the nucellus becomes a single layer of cells except on the ventral side at the crease where it forms a proliferation of cells projecting into the endosperm. Outside the nucellus is the testa, both these tissues having an outer cuticular layer except at the crease (Cochrane and Duffus, 1979). The inner cells of the pericarp contain chloroplasts and are one cell thick in oats, rye and wheat, and 2 to 3 cells thick in barley. The outer pericarp is transparent throughout development, contains small starch granules and has an external cuticular layer. Stomata are present on the ventral side near the apex of the caryopsis.

Nutrients are thought to enter the developing endosperm *via* the vascular bundle in the pericarp and pass through the gap in the cuticular layers into the nucellar projection and then into the endosperm. Sucrose is generally considered to be the major carbon source but other carbon-containing nutrients which may reach the developing endosperm include amino acids such as glutamate/glutamine and aspartate/asparagine (Duffus and Rosie, 1978).

The embryo thus develops at the edge of a tissue which for the greater part of grain development is actively respiring and synthesising carbohydrate (Duffus, 1979). Oxygen is probably in short supply, and although it is produced up to a certain stage in development by photosynthesis in the pericarp, limited amounts only may penetrate the cuticular barriers to reach the growing embryo.

CONTROL OF EMBRYO DEVELOPMENT AND GERMINATION

Immature embryos from many species when excised and grown in culture are capable of precocious germination, that is, where normal maturation is prevented. Precocious germination can be defined as germination which takes place before the embryo has reached full size and maturity. Precocious germination in barley can be suppressed and 'normal' embryonic development can continue followed by germination if the sucrose concentration of the medium is increased to around 0.35M (Cameron-Mills and Duffus, 1980) or if the medium contains abscisic acid (ABA) (see Walton, 1980). Such observations suggest that, during normal development *in vivo*, the information required to suppress precocious germination comes from the environment external to the embryo rather than from the embryo itself. ABA is considered to exert its effect at the level of protein synthesis and in particular the inhibition of gibberellic acid (GA) - triggered synthesis of α-amylase in barley aleurone layers (see Walton, 1980).

In the early stages of grain development, endosperm sucrose levels are high (Cameron-Mills and Duffus, 1980) although it is likely that at this stage the sucrose supply limits starch accumulation (Jenner, 1982). At the same time ABA levels in the caryopsis are also high (King *et al.*, 1979; Slominski *et al.*, 1979). If these observations relate to the embryo environment then they should combine to prevent germination. In the middle to later stages of grain development, endosperm sucrose content falls somewhat although it is likely that supplies are in excess of requirements (Jenner, 1982). In maize embryos, sucrose is the dominant sugar present throughout maturation and concentrations (W/W) are much higher than in endosperms of the same age (Gentinetta *et al.*, 1979). ABA levels fall markedly over the later stages of maturation (King *et al.*, 1979) and indeed are low in most non-dormant seeds at the end of maturation (Walton, 1980). Unfortunately the amounts present in the embryos are rarely quoted (but see Radley, 1979). High levels of ABA in immature caryopses might serve to prevent precocious germination. Low levels at maturity do not necessarily mean that ABA is not involved *in vivo* since it may be that the mature embryo has an increased sensitivity to the inhibitory effects of ABA. In a cultivar of *Triticale*, where 20% of kernels may sprout by 55 days after anthesis and α-amylase levels increase dramatically, levels of ABA were no lower than those found in 'normal' cultivars (King *et al.*, 1979). Again this does not exclude ABA involvement since these kernels may be less sensitive to ABA than are other kernels.

Isolated immature barley embryos can take up sucrose, and the rate of uptake increases with embryo development. From the kinetics of sucrose uptake it can be deduced (Cameron-Mills and Duffus, 1979) that active transport predominates at sucrose concentrations of 50 mM which approximates to those in the endosperm at the middle to later stages of development. At higher concentrations passive diffusion makes an increasing contribution to uptake. Interestingly most of the sucrose taken up accumulates in a pool of free sucrose and only some 20% is utilised in biosynthetic processes. The levels are presumably regulated by UDP- and ADP-dependent sucrose synthases and invertase. Of these, invertase remains active in barley embryos at maturity (Duffus and Rosie, 1975). Thus it may be that the principal function of sucrose entering the embryo is to maintain a high osmotic pressure within the embryo cells. Any decrease in its rate of uptake or increase in the rates of utilisation, for example in respiration, might result in germination.

Crouch and Sussex (1981) working with developing *Brassica napus* embryos *in vivo* and *in vitro* have suggested that ABA and sucrose may be involved in preventing precocious germination by allowing embryogeny to continue. Embryogeny was monitored by comparing accumulation of storage proteins in culture with normal accumulation in seeds. Precocious germination could be induced by removing embryos from the developing seeds at 27 days after anthesis, just as storage protein was beginning to accumulate, and placing them in culture on a basal medium. Amino acid incorporation into one of the two major storage proteins (12S) decreased. If ABA (10^{-6}M) was included in the medium, germination was prevented, amino acid incorporation increased and eventually the accumulation rate equalled that observed *in vivo*. If the seeds were transferred to a solution of high osmotic pressure such as 0.29M sucrose, precocious germination was also inhibited but there was a lag phase before 12S protein synthesis rates equalled those on ABA media. Again this suggests that the maternal environment controls embryogeny, possibly mediated by ABA or sucrose.

In cultured axes of *Phaseolus vulgaris* ABA has a negligible effect on embryogeny as measured by phaseolin synthesis (Long *et al.*, 1981). In this case it seems unlikely that the maternal environment is responsible for preventing precocious germination. The behaviour of the axes might be explained by their having an internal supply of ABA or a sufficiently high internal osmotic pressure which would prevent germination.

Further evidence supporting the role of ABA in the prevention of precocious germinations comes from the work of Smith *et al.*,(1978) who showed that maize mutants which germinate precociously have a decreased sensitivity to ABA inhibition or have reduced levels of ABA.

In contrast Berrie *et al.*,(1979) have shown that ABA levels were higher in *Avena sativa*, a species which at most shows a brief post-harvest dormancy, than in *Avena fatua* (wild oat), a species which is appreciably dormant. Levels of ABA in *Avena fatua* at harvest were similar to those present at germination. Again if endogenous ABA does prevent germination then it must be the sensitivity of the receptor that is important rather than absolute levels of inhibitor. The same authors have suggested a positive correlation between dormancy and grain content of volatile fatty acids and that their evaporation leads to increased germination. More recently Dunwell (1981) has suggested that the endogenous ABA levels cannot be the factor controlling post-harvest dormancy since in one cultivar of barley, BR 76R 23-1, known to possess post-harvest dormancy, none of the grains analysed post-harvest contained detectable ABA. Unfortunately no evidence was presented to indicate the sensitivity of the method used. In these, as with all similar correlations, the extent of variation throughout the whole range of species is unknown and so the validity of the conclusions is open to question.

Interrelationship of GA, ABA and respiration

The role of GA in the control of α-amylase synthesis during germination has been the subject of much investigation and its function during malting in barley is fairly well understood (Palmer, 1980). It is possible that germination may be controlled by a balance between endogenous ABA and GA levels (Radley, 1979) and/or water stress (high sucrose concentration). For example Mozer (1980), using isolated 'aleurone' tissue, has shown that, when applied separately, ABA and GA induce the formation of new translatable m-RNAs. In the case of GA, large amounts of α-amylase are produced. When both hormones are present, ABA prevents the translation of the m-RNAs but does not inhibit their induction by GA.

Clearly however, metabolic activity must be present from the earliest stages of germination to supply both energy and intermediates for hormone and protein synthesis. These may be derived from normal aerobic processes which are active during the mid-late stages of grain development. As the oxygen supply decreases, the enzymes of the pentose phosphate pathway (PPP) remain active and those characteristic of aerobic metabolism decrease in activity (Duffus and Rosie, 1977). Since PPP activity is also dominant during early germination (see Palmer, 1980) when oxygen may also be in short supply, it may be that the energy and intermediates required for pre-harvest sprouting, if not precocious germination, is derived from the PPP. There is no evidence that PPP intermediates are involved in the initiation of germination, that is in the synthesis of GA or the degradation of ABA. It may be of course, that by decreasing the sucrose concentration, respiratory activity reduces the osmotic pressure within the embryo cells and this induces germination as observed above with isolated cultured embryos.

CONCLUSIONS

This review has shown how biochemical events in the developing grain may influence the development, maturation and germination of the embryo. The mechanism whereby such influence may be exerted is not clear since the direct source of nutrients and other substances required for embryogenesis remains inknown. There is some evidence to suggest that ABA, water stress and GA may be involved in the control of germination. The mechanisms concerned in regulating the production of GA, the synthesis and degradation of ABA and the intracellular levels of sucrose remain a mystery.

REFERENCES

Ahluwalia, B. 1980. Mineral elements and embryo development in barley. PhD Thesis. University of Edinburgh.

Berrie, A.M.M., Buller, D., Don, R., and Parker, W. 1979. Possible role of volatile fatty acids and abscisic acid in the dormancy of oats. Plant Physiol. 63, 758-764.

Cameron-Mills, V. and Duffus, C.M. 1979. Sucrose transport in isolated immature barley embryos. Ann. Bot. 43, 559-569.

Cameron-Mills, V. and Duffus, C.M. 1980. The influence of nutrition on embryo development and germination. Cereal Res. Commun. 8, 143-149.

Cochrane, M.P. and Duffus, C.M. 1979. Morphology and ultrastructure of immature cereal grains in relation to transport. Ann. Bot. 44, 67-72.

Crouch, M.L. and Sussex, I.M. 1981. Development and storage - protein synthesis in Brassica napus L. embryos in vivo and in vitro. Planta 153, 64-74.

Duffus, C.M. and Rosie, R. 1975. Biochemical changes during embryogeny in Hordeum distichum. Phytochem. 14, 319-323.

Duffus, C.M. and Rosie, R. 1977. Carbohydrate oxidation in developing barley endosperm. New Phytol. 78, 391-395.

Duffus, C.M., and Rosie, R. 1978. Metabolism of ammonium ion and glutamate in relation to nitrogen supply and utilisation during grain development in barley. Plant Physiol. 61, 570-574.

Duffus, C.M. 1979. Carbohydrate metabolism and cereal grain development. In: Recent Advances in the Biochemistry of Cereals (D.L. Laidman and R.G. Wyn-Jones eds.) Academic Press pp. 209-238.

Duffus, C.M. and Cochrane, M.P. (1982). Carbohydrate metabolism during cereal grain development. In: The Physiology and Biochemistry of Seed Development, Dormancy and Germination. A.A. Khan, (ed.) Elsevier.

Dunwell, J.M. (1981). Dormancy and germination in embryos of Hordeum vulgare L. - Effect of dissection, incubation temperature and hormone application. Ann. Bot. 203, 203-213.

Gentinetta, E., Zambello, M. and Salamini, F. 1979. Free sugars in developing maize grain. Cereal Chem. 56, 81-83.

Jenner, C.F. 1982. Storage of starch. In: Encyclopaedia of Plant Physiology (N.S.) Vol. 13B. Plant Carbohydrates I. (W. Tanner and F.A. Loewus, eds.) Springer-Verlag.

King, R.W., Salminen, S.O., Hill, R.D. and Higgins, T.J.V. 1979. Abscisic acid and gibberellin action in developing kernels of Triticale (cv. 6 A 190). Planta 146, 249-255.

Long, S.R., Dale, R.M.K. and Sussex, I.M. 1981. Maturation and germination of Phaseolus vulgaris embryonic axes in culture. Planta 153, 405-415.

Mozer, T.J. 1980. Control of protein synthesis in barley aleurone layers by the plant hormone gibberellic acid and abscisic acid. Cell 20, 479-485.

Norstog, K. 1972. Early development of the barley embryo: fine structure. Amer. J. Bot. 59, 123-132.

Palmer, G.H. 1980. The morphology and physiology of malting barleys. In: Cereals for Food and Beverages. Academic Press, Inc. pp. 301-338.

Radley, M., 1979. The role of gibberellin, abscisic acid and auxin in the regulation of developing wheat grains. J. Exp. Botany 30, 381-389.

Slominski, B., Rejowski, A. and Nowak, J. 1979. Abscisic acid and gibberellic acid contents in ripening barley seeds. Physiol. Plant. 45, 167-169.

Smith, J.D., McDaniel, S. and Lively, S. 1978. Regulation of embryo growth by abscisic acid *in vitro*. Maize Genet. Newsletter 52 107-108.

Walton, D.C. 1980. Biochemistry and physiology of abscisic acid. Ann. Rev. Plant Physiol. 31, 453-489.

Separation of Wheat Alpha-Amylase Isoenzymes by Chromatofocusing

B. A. Marchylo and J. E. Kruger,
*Grain Research Laboratory, Canadian Grain
Commission, Winnipeg, Manitoba, Canada*

SUMMARY

Chromatofocusing was used to resolve germinated wheat α-amylase isoenzymes. Optimum resolution and recovery of wheat α-amylase isoenzymes was achieved with a polybuffer ^{TM} 74^2 dilution factor of 10x, a buffer pH interval of 7.4 to 4.8, a column bed height of 45 cm, a flow rate of 0.5 mL/min, eluent buffer containing 10^{-4} M $CaCl_2$ and a temperature of 4°C. A maximum of six peaks were resolved under these conditions. The first 3 peaks eluted were composed of GIII α-amylase isoenzymes with only a trace of GII isoenzymes. The remaining 3 peaks were composed of GI isoenzymes. The usefulness of chromatofocusing for separation of germinated wheat α-amylase isoenzymes was assessed.

INTRODUCTION

The polymorphic nature of germinated and immature wheat α-amylase has been demonstrated in numerous studies (Marchylo et al. 1981, 1980a,b, 1976; Daussant et al. 1980; Kruger 1972a,b; Olered and Jonsson 1970; Kruger and Tkachuk 1969). Chromatofocusing is a recently developed column chromatographic technique (Sluyterman and Elgersma 1978; Sluyterman and Wijdenes 1978, 1981a,b; Pharmacia Fine Chemicals 1980-1.) which separates proteins on the basis of isoelectric pH. It has been reported that this technique offers high resolution with peak widths ranging down to 0.04-0.05 pH unit (Pharmacia Fine Chemicals 1980-1). The purpose of this study was to assess the ability of chromatofocusing (using the commercially available Pharmacia Fine Chemicals Chromatofocusing System, Polybuffer ^{TM} 74^2 and PBE ^{TM} 94^2) to resolve germinated wheat α-amylase isoenzymes.

MATERIALS AND METHODS

Germinated hard red spring (HRS) wheat cv. Neepawa was prepared by first steeping 15 g of mature wheat for 2 hrs at room temperature. The steeped wheat then was placed in a petri dish on water-saturated blotting paper and germinated for 5 days at 18-20°C and at

Paper No. 490, Canadian Grain Commission, Grain Research Laboratory, 1404-303 Main Street, Winnipeg, Manitoba R3C 3G8.

about 96% relative humidity. The germinated wheat was bench dried before use.

Extraction Procedure. Unless stated otherwise, 10 kernels of 5 day germinated wheat were homogenized with a pestle in a mortar containing 1 g of sand and 10 mL of cold buffer (0.025 M imidazole - HCl, pH 7.4). The extract was centrifuged twice at 23,600 x g at 4°C for 15 min.

Preparation of Heat-treated Extracts. Fifty 5 day germinated wheat kernels were homogenized with a pestle in a mortar containing 1 g of sand and 5 mL of cold buffer (0.2 M sodium acetate, pH 5.5, containing 10^{-3} M $CaCl_2$). The extract was poured into a VirTis homogenizing flask along with 10 mL of buffer. This was mixed with a VirTis Model 23 homogenizer for 1 min at a speed setting of 20. The extract was centrifuged at 23,600 x g at 4°C for 10 min. The supernatant was heat-treated at 70°C for 15 min. It then was centrifuged twice at 23,600 x g at 4°C for 15 min. The supernatant was dialyzed overnight against 3 L of buffer (0.025 M imidazole, pH 7.4).

Column Preparation. Polybuffer exchanger 94 (PBE) was equilibrated and packed as recommended by the manufacturer (Pharmacia Fine Chemicals 1980-1). The start buffer used was 0.025 M imidazole - HCl, pH 7.4. Pharmacia K9/30 and K9/60 columns were used to prepare columns with bed heights of about 17 cm or 45 cm. Columns were packed at flow rates of 2.5 mL/min. Two to four cm of Sephadex[TM] [2] G-25 coarse was layered on top of the bed to ensure even sample application.

Sample Application and Elution. Samples were applied by gravity (Pharmacia Fine Chemicals 1980-1). Following sample application, eluent buffer (Polybuffer 74 diluted 10x with distilled water, pH 4.3 or 4.8 as required, containing 10^{-4} M $CaCl_2$) was pumped through the column at 0.5 mL/min. Fractions (2 mL) were collected with an LKB UltroRac fraction collector (LKB Producter AB, Bromma, Sweden) equipped with a drop counter. Columns were regenerated as described by the manufacturer (Pharmacia 1980-1).

α-Amylase Activity Analysis. Total α-amylase activity per fraction was determined, as described by Kruger and Tipples (1981), with a Perkin-Elmer Model 191 Grain Amylase Analyzer.

Isoelectric Focusing. Prepared thin-layer polyacrylamide gel plates (Ampholine PAG plate kits, pH 3.5-9.5, LKB Producter AB, Bromma, Sweden) were used for isoelectric focusing. Isoelectric focusing was carried out with a Desaga thin layer electrophoresis apparatus cooled to 2°C. Weight (450 g) was placed across the electrodes to facilitate contact between gel, wick and electrode. Constant power (10 W) was maintained throughout a 2 hr run by an ISCO

[2] Polybuffer 74, PBE and Sephadex are the exclusive trademarks of Pharmacia Fine Chemicals AB, Uppsala, Sweden.

Model 494 electrophoresis power supply. The pH gradient across the PAG plate was determined with a Multiphor pH surface electrode (LKB Producter, Bromma, Sweden).

Detection of α-Amylase Isoenzymes. α-Amylase isoenzymes were detected using β-limit dextrin plates previously described (Marchylo et al, 1980a; MacGregor et al, 1974). Incubation time for the β-limit dextrin-polyacrylamide sandwich was 15-20 min for eluent fractions and 5-7 min for original extracts.

RESULTS

Optimum chromatofocusing conditions for resolution and recovery of wheat α-amylase isoenzymes were determined by varying polybuffer 74 dilution factor, buffer pH interval, column bed height, flow rate, CaCl$_2$ concentration and temperature.

Resolution of wheat α-amylase isoenzymes increased with decreasing pH gradient slope. pH gradient slope decreased as the pH interval between start and eluent buffers decreased and as column bed height increased. A comparison of chromatograms (Fig. 1 and 2A) shows that a decrease in pH interval (pH 7.4-4.3 to 7.4-4.8) and an increase in column bed height (17 cm to 45 cm) results in a shallower pH gradient and greater resolution. Increased resolution is indicated by greater peak separation and by an increase in the number of resolved peaks. Increasing the polybuffer 74 dilution factor also decreased pH gradient slope. However, a dilution factor of 10x was chosen as optimum since it gave the best compromise between resolution, run time and cost.

CaCl$_2$ was included in the eluent buffer since α-amylase requires Ca^{++} for stability (Fischer and Stein, 1960). The inclusion of CaCl$_2$ did not affect the pH gradient, however concentrations higher than 10^{-4} M led to higher elution pH's and to a substantial loss in resolution. For this reason 10^{-4} M CaCl$_2$ was used routinely in chromatofocusing analyses.

The effect of flow rate on resolution was assessed and a rate of 0.5mL/min was chosen as optimal for resolution and running time.

Enzyme recoveries for analyses carried out at room temperature were substantially lower than recoveries at 4°C. Duplicate analyses carried out under identical chromatographic conditions (17 cm bed height; polybuffer 74 dilution factor of 10x; 10^{-4} M CaCl$_2$; pH gradient interval of 7.4-4.3; flow rate of 0.5 mL/min) gave average recoveries of 84.5% at 4°C but only 17.5% at room temperature. A somewhat lower average recovery of 75.1% was achieved at 4°C when a longer column bed height (45 cm) and a smaller pH gradient interval (pH 7.4-4.8) was used.

pH gradients generated by the chromatofocusing procedure were curvilinear (Fig. 1, 2A,B). Although minor variations in pH gradients were found between chromatographic runs, peaks focused at reproducible elution pH's. The average elution pH's for the five peaks

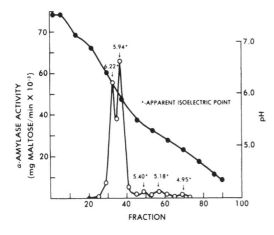

Figure 1. Separation of germinated wheat alpha-amylase by chromatofocusing (open symbols). pH gradient obtained during chromatofocusing (closed symbols). Column: K9/30. Bed height: 17 cm. Elution conditions: Start-buffer 0.025M imidazole-HCl, pH 7.4; Eluent buffer Polybuffer 74 diluted 10x containing 10^{-4} M CaCl$_2$, pH 4.3; Flow rate 0.5 mL/min; Temperature 4°C; Fraction volume 2 mL.

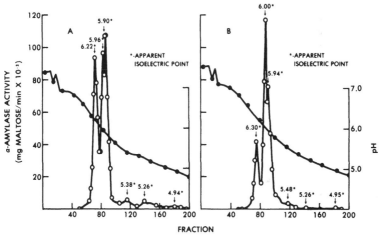

Figure 2. A. Separation of germinated wheat α-amylase by chromato-focusing (open symbols). B. Separation of heat-treated germinated wheat α-amylase by chromatofocusing (open symbols). A and B. pH gradient obtained during chromatofocusing (closed symbols). Column: K9/60. Bed height 45 cm. Elution conditions: Start buffer 0.025M imidazole-HCl, pH 7.4; Eluent buffer Polybuffer 74 diluted 10x containing 10^{-4}M CaCl$_2$, pH 4.8; Flow rate 0.5 mL/min; Temperature 4°C; Fraction volume 2 mL.

resolved in Fig. 2 were 6.22 ± 0.03, 5.94 ± 0.03, 5.38 ± 0.03, 5.22 ± 0.04 and 4.94 ± 0.03 for 6 separate chromatofocusing runs. Resolution was affected somewhat by minor variations in pH gradient. For example, the two closely spaced peaks with elution pH's of 5.96 and 5.90 were not always resolved (Fig. 2A,B) due to such variations. Similarly, relative peak heights could vary between runs. Heat treatment at 70°C did not alter the number of peaks resolved by chromatofocusing but did result in significant changes in relative peak heights (Fig. 2B).

Eluent fractions were selected and analyzed by polyacrylamide gel isoelectric focusing (PAGIEF) to determine the identity of α-amylase isoenzymes resolved by chromatofocusing. Isoenzyme composition of eluent fractions were compared to the isoenzyme composition of the original α-amylase extract (Fig. 3). For discussion purposes, α-amylase isoenzymes were divided into 3 groups as described previously (Marchylo et al, 1980a). Comparable results were

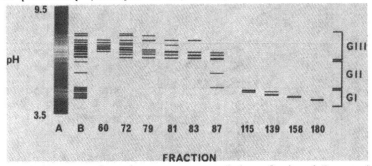

Figure 3. The α-amylase isoenzyme composition of: A and B, germinated wheat extract; Fractions 60-180, eluent fractions obtained by chromatofocusing as illustrated in Fig. 2A.

obtained for each chromatofocusing run (Fig. 1; 2A,B) with the exception that more overlap of isoenzymes was evident for the chromatogram depicted in Fig. 1. As illustrated in Fig. 3, all but 2 GII isoenzymes present in the original extract were recovered following chromatofocusing. Four major GIII and 3 major GI isoenzymes, as assessed by visual band intensity, were resolved by PAGIEF analysis of the original wheat extract (Fig. 3A,B). The GIII isoenzymes had pI's of 6.5, 6.4, 6.3 and 6.2 while the 3 GI isoenzymes had pI's of 5.10, 4.95 and 4.75.

PAGIEF analysis of the first 3 peaks eluted revealed that each contained a mixture of α-amylase isoenzymes (Fig. 3). The first peak eluted (elution pH 6.22, Fig. 2A) contained only GIII α-amylase isoenzymes. The GIII isoenzyme composition within this peak changed across the peak as illustrated in Fig. 3 (Fractions 60-79). One major GIII isoenzyme was focused in this peak. This isoenzyme increased in activity to a maximum in Fraction 72 and then decreased. The 3 remaining major GIII isoenzymes also were present. However, re-chromatofocusing of Fraction 72 indicated that these 3 major isoenzymes were present due to overlap with the next two peaks (elution pH 5.96 and 5.90). The minor GIII α-amylase isoenzymes did not elute strictly in order of decreasing pI in going from Fraction 54-79. For example, minor isoenzymes were present in Fraction 72, which had a higher pI than the highest pI isoenzyme present in Fraction 60 (Fig. 3).

The closely spaced second and third eluting peaks (elution pH 5.96 and 5.90) mainly contained GIII isoenzymes with a trace of GII isoenzymes detectable (Fraction 87). Three major GIII isoenzymes were eluted in order of decreasing pI within this peak (Fractions 79-87,Fig. 3). Fewer minor GIII isoenzymes were eluted and as in the first peak these did not elute strictly according to decreasing pI.

The last three peaks eluted (elution pH 5.38, 5.26 and 4.94, Fig. 2A) contained GI isoenzymes. These eluted according to decreasing pI. Resolution of the GI isoenzymes was substantially better than resolution of the GIII isoenzymes as indicated by less overlap of isoenzymes between peaks (Fig. 3).

DISCUSSION

Optimum resolution and recovery of wheat α-amylase isoenzymes, by chromatofocusing, was achieved with a polybuffer 74 dilution factor of 10x, a buffer pH interval of 7.4-4.8, a column bed height of 45 cm, a flow rate of 0.5 mL/min, eluent buffer containing 10^{-4} M $CaCl_2$ and a running temperature of 4°C. Under these conditions average enzyme recoveries of about 75% could be achieved. Substantially lower recoveries of about 17% were achieved when analyses were carried out at room temperature. Wheat α-amylase isoenzymes are known to be relatively stable at temperatures below 40°C (Tkachuk and Kruger 1974; Marchylo et al. 1974), therefore it is surprising that such a substantial loss in activity is observed at room temperature. Possibly, the α-amylase interaction with PBE makes it more susceptible to heat denaturation.

Resolution and elution pH were affected strongly by $CaCl_2$ concentrations higher than 10^{-4} M. This effect may have been due to increased ionic strength. It is known that elution pH is closest to the pI of the protein when ionic strength is low (Pharmacia Fine Chemicals 1980-1). A maximum of 6 peaks were resolved by chromatofocusing analysis of wheat α-amylase (Fig. 2A). Four major GIII α-amylase isoenzymes were located in the first three peaks eluted during chromatofocusing. The first peak eluted (elution pH 6.22, Fig. 2A) contained one major GIII isoenzyme, (pI 6.50). The next two peaks (elution pH 5.96 and 5.90) contained the remaining 3 major GIII isoenzymes (pI 6.4, 6.3 and 6.2). Comparison of pI's obtained for these isoenzymes by PAGIEF and by chromatofocusing is difficult because of the mixture of α-amylase isoenzymes present. However, since the first peak eluted contains only one major GIII isoenzyme it would appear safe to make a comparison. The result obtained by chromatofocusing was 0.28 pH units lower than that obtained by PAGIEF. This is not surprising since it has been shown that in chromatofocusing a protein will elute at a pH somewhat lower than its pI (Pharmacia Fine Chemicals 1980-1; Sluyterman and Elgersma 1978; Sluyterman and Wijdenes 1978). For this reason pI's obtained by chromatofocusing were designated as apparent pI's (pI app).

Since the remaining 3 major GIII isoenzymes are not completely resolved from each other during chromatofocusing (Fig.2A), a comparison of pI's and pI's app is more difficult. However, it is evident that the pI app values also are lower than the corresponding pI's.

It is surprising that chromatofocusing was able to resolve one of the 4 major GIII isoenzymes into a discrete peak (pI app 6.22) whereas the remaining 3 major GIII isoenzymes essentially were eluted together (peaks pI app 5.96 and 5.90, Fig. 2A). PAGIEF analysis showed that the pI difference between each successive major GIII isoenzyme (Fig. 3A) was the same. Therefore, these isoenzymes should have been equally resolved during chromatofocusing. This suggests that physio-chemical differences other than pI may influence the resolution of the 4 major GIII isoenzymes by chromatofocusing. It is of note that in previous work (Kruger and Tkachuk 1969; Tkachuk and Kruger 1974) 4 major germinated wheat α-amylase isoenzymes were resolved by ion-exchange chromatography.

Minor GIII isoenzymes did not elute in sequential order of descending pI but appeared instead to elute as a group. Such behavior is abnormal since chromatofocusing should separate proteins in order of descending pI. This raises the question as to whether or not these isoenzymes are artifacts. This is difficult to determine since both PAGIEF and chromatofocusing resolve α-amylase isoenzymes based upon their pI's. Therefore, artifacts could result from either procedure. It should be realized that minor GIII isoenzymes are not necessarily artifacts of the separatory procedures themselves. Possibly other protein(s) focusing at the same pI as α-amylase isoenzymes combine with the isoenzymes to produce the minor GIII isoenzymes. Analysis of purified wheat α-amylase by chromatofocusing and PAGIEF may give further information concerning these minor GIII components.

The GI family of α-amylase isoenzymes (3 major and 3 minor isoenzymes, Fig. 3A,B) eluted in 3 broad peaks (pI app 5.38, 5.26 and 4.94; Fig. 2A) in order of decreasing pI. The 3 major GI isoenzymes were resolved from each other with little overlap of peaks being observed. Minor GI isoenzymes also were present in peaks pI app 5.38 and 5.26. The inability of chromatofocusing to resolve these isoenzymes undoubtedly was a reflection of extremely small differences in pI. A comparison of pI's (5.10, 4.95, 4.75) and pI's app. (5.38, 5.26 and 4.94) for the major GI isoenzymes demonstrates that the pI's app were higher. This behavior, which is opposite to that obtained for the major GIII isoenzymes, may result from increasing ionic strength during chromatofocusing or from the type of titration curve exhibited by the GI isoenzymes. It has been shown that these factors can displace proteins during chromatofocusing (Pharmacia Fine Chemicals 1980-1).

Although heat treatment of wheat α-amylase did not affect the number of peaks resolved by chromatofocusing, it did change relative peak heights (Fig. 2B). Peaks pI app 6.30 and 5.94 (GIII), 5.48, 5.26 and 4.95 (GI) decreased substantially following the heat treatment procedure. The decrease in GI isoenzyme relative activity is consistent with a previous study which showed that GI-type isoenzymes present in immature wheat (Marchylo et al, 1974) are more heat labile than GIII isoenzymes of germinated wheat (Tkachuk and Kruger 1974). However, it also appears that differences in heat stability are present within the GIII family of isoenzymes.

An overall assessment of the chromatofocusing technique indicates that this pocedure can be used to good advantage in the separation and study of the germinated wheat α-amylase isoenzyme system. Chromatofocusing has the following advantages which make it suitable for study of this system:

1) Isoenzymes elute in a relatively small volume of eluent buffer. This simplifies further analysis since large volumes do not have to be concentrated.

2) The complete system of germinated wheat α-amylase isoenzymes can be separated on one column in one run. This is unlike ion-exchange column chromatography where separate conditions and runs are required to resolve the GIII and GI isoenzymes (Marchylo et al. 1974; Kruger and Tkachuk 1969).

3) Chromatofocusing is a simple, fast and relatively inexpensive procedure.

4) Chromatofocusing can separate quantitatively the α-amylase isoenzymes whereas PAGIEF produces a qualitative separation.

5) Resolution of the GI isoenzymes is excellent and it appears to be superior to the resolution obtained by ion exchange column chromatography (Marchylo et al. 1974).

One disadvantage of chromatofocusing appears to be an inability to resolve all the major GIII isoenzymes. However, this inability may provide valuable information about differences between these isoenzymes. The elution behavior of the minor GIII isoenzymes is puzzling. Further studies into the nature of these isoenzymes are required.

ACKNOWLEDGEMENTS

The authors would like to thank Sandy Gowan and Donna Levich for their excellent technical assistance and Dr. A.W. MacGregor for his helpful comments and discussion.

REFERENCES

Daussant, J., Mayer, C., and Renard, H.A. 1980. Immunochemistry of Cereal α-Amylases in Studies Related to Seed Maturation and Germination. Cereal Res. Commun. 8:49-60.

Fischer, E.H., and Stein, E.A. 1960. Alpha-amylases. In: The Enzymes, ed. by P. Boyer, H. Lardy, and K. Myrbach, Vol. 4, p.313-343 Academic Press, New York.

Kruger, J.E. 1972a. Changes in the Amylases of Hard Red Spring Wheat During Growth and Maturation. Cereal Chem. 49:379-390.

Kruger, J.E. 1972b. Changes in the Amylases of Hard Red Spring Wheat During Germination. Cereal Chem. 49:391-398.

Kruger, J.E., and Tipples, K.H. 1981. Modified Procedure for Use of the Perkin-Elmer Model 191 Grain Amylase Analyzer in Determining Low Levels of α-Amylase in Wheats and Flours. Cereal Chem. 58:271-274.

Kruger, J.E., and Tkachuk, R. 1969. Wheat Alpha-amylases. I. Isolation. Cereal Chem. 46:219-226.

MacGregor, A.W., Thompson, R.G., and Meredith, W.O.S. 1974. α-Amylase from Immature Barley: Purification and Properties. J. Inst. Brew. 80:181-187.

Marchylo, B., Kruger, J.E., and Irvine, G.N. 1976. α-Amylase from Immature Hard Red Spring Wheat. I. Purification and Some Chemical and Physical Properties. Cereal Chem. 53:157-173.

Marchylo, B.A., LaCroix, L.J., and Kruger, J.E. 1980a. α-Amylase Isoenzymes in Canadian Wheat Cultivars during Kernel Growth and Maturation. Can. J. Plant Sci. 60:433-443.

Marchylo, B.A., LaCroix, L.J., and Kruger, J.E. 1980b. The Synthesis of α-Amylase in Specific Tissues of the Immature Wheat Kernel. Cereal Res. Commun. 8:61-68.

Marchylo, B.A., LaCroix, L.J., and Kruger, J.E. 1981. α-Amylase Synthesis in Wheat Kernels as Influenced by the Seed Coat. Plant Physiol. 67:89-91.

Olered, R., and Jonsson, G. 1970. Electrophoretic Studies of α-Amylase in Wheat. II. J. Sci. Fd. Agric. 21:385-392.

Pharmacia Fine Chemicals. 1980-1. Chromatofocusing with Polybuffer TM and PBE .

Sluyterman, L.A.AE., and Elgersma, O. 1978. Chromatofocusing: Isoelectric Focusing on Ion-exchange Columns I. General Principles. J. Chromatogr. 150:17-30.

Sluyterman, L.A.AE., and Wijdenes, J. 1978. Chromatofocusing: Isoelectric Focusing on Ion-exchange Columns. II. Experimental Verification. J. Chromatogr. 150:31-44.

Sluyterman, L.A.AE., and Wijdenes, J. 1981a. Chromatofocusing. III. The Properties of a DEAE-agarose Anion Exchanger and its Suitability for Protein Separations. J. Chromatogr. 206:429-440.

Sluyterman, L.A.AE., and Wijdenes, J. 1981b. Chromatofocusing. IV. Properties of an Agarose Polyethyleneimine Ion Exchanger and its Suitability for Protein Separations. J. Chromatogr. 206: 441-447.

Tkachuk, R., and Kruger, J.E. 1974. Wheat α-Amylases. II. Physical Characterization. Cereal Chem. 51:508-529.

Alpha-Amylase Genes in Wheat

M. D. Gale, *Plant Breeding Institute, Cambridge, England*

INTRODUCTION

At previous Preharvest Sprouting Symposia, a considerable amount of information concerning the chemical and developmental aspects of α-amylase production in wheat has been presented. The various forms of the enzyme are, of course, products of several structural genes, and genetic analysis can reveal additional information concerning the relationships between the various isozymes.

The initial identification of the chromosomal locations of the α-amylase structural genes was made by Nishikawa and Nobuhara (1971). They identified the involvement of the three chromosomes in group 6, 6A, 6B and 6D in control of some of the isozyme bands with more basic isoelectric points (pIs) and 7A, 7B and 7D in the control of some of the isozymes with acidic pIs. They also noted some varietal differences in isozyme patterns.

The development of improved flat bed isoelectric focussing techniques and the need for genetic markers in wheat has prompted further analysis of these two triplicate sets of genes, now known as α-Amy-1 and α-Amy-2, by Nishikawa et al. (1981) and Gale et al. (1982). The gene products of these two gene series are, in wheat genetics, referred to as α-AMY1 and α-AMY2.

GENETIC ANALYSIS

The most common method of ascertaining the chromosomal locations of enzyme structural genes in wheat is nullisomic analysis, usually by use of the Chinese Spring (CS) nullisomic-tetrasomic (NT) and ditelocentric (DT) stocks developed by E.R. Sears at Columbia, Missouri. NTs are genotypes in which the complete absence of a single chromosome pair is compensated by the addition of an extra pair of one of the other chromosomes in the same homoeologous group. DTs are genotypes in which one of the 21 pairs of homologues lacks an entire chromosome arm. With both NTs and DTs structural genes on the chromosome or arm not present can be identified by the absence of gene products relative to the normal euploid CS phenotype.

FIGURE 1

α-Amylase (α-AMY2) isozymes produced in the Chinese Spring nullisomic-tetrasomic lines for homoeologous group 7.

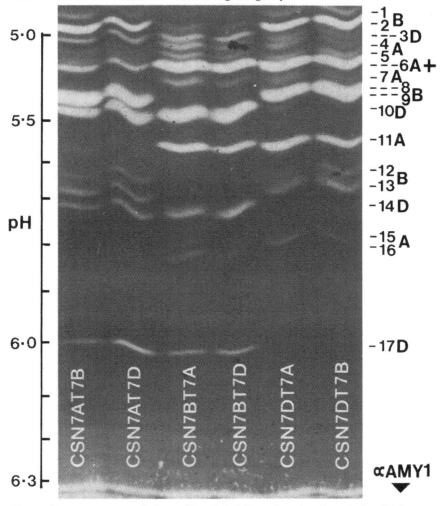

Note. Enzyme extracted from distal half grains incubated for 72 h at
25°C in 10 mM acetate buffer (pH 4.8), 10 mM CaCl₂ and 10 μM
GA₃. Gel LKB pH range 4 – 6.5. Isoelectric focussing technique as described by Sargeant and Walker (1978). N7AT7B
CS nullisomic 7A – tetrosomic 7B etc.

Figure 1 shows the α-amylase isozymes produced by distal half-
grains of the CS Group 7 NTs. The 17 bands in the lower pI region

of the gel can almost all be seen to be the products of a single group 7 chromosome. For example chromosome 7B is responsible for the control of bands 1, 2, 8, 9, 12 and 13.

FIGURE 2

αAMY2 isozymes produced in Lutescens 62 and the Group 7 Chinese Spring (Lutescens 62) intervarietal chromosome substitution lines.

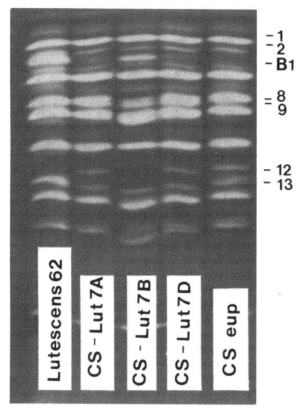

Observation of a range of varieties reveals considerable varia- tion in isozyme patterns. Differences that involve the loss of one of the CS bands can be assumed to be due to allelic variation on the chromosome already identified by nullisomic analysis. Differences that involve isozymes not produced by CS require the use of various aneuploid techniques to identify the chromosomal source of the alle- lic difference. Figure 2 shows such variation among the αAMY2 iso- zymes of Lutescens 62 and CS. Lutescens 62 displays an extra band at the same position as CS band 4 and lacks band 12. Analysis of the intervarietal chromosome substitution lines of Lutescens 62 into CS developed by C.N. Law and A.J. Worland at the Plant Breeding Ins- titute confirm that the variation is associated with chromosome 7B.

FIGURE 3

α AMY1 isozymes produced by <u>Triticum spelta</u> and the group 6 Chinese Spring (<u>T. spelta</u>) intervarietal substitution lines.

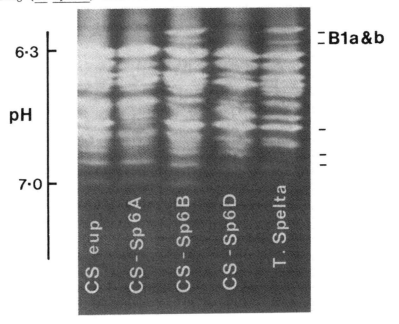

This analysis does not, however, allow any deductions to be made concerning the arrangement of the structural gene units controlling production of these bands within the chromosome. This information can be gained from analysis of segregating allelic differences.

The variety Hope shows a similar phenotype to Lutescens 62 and the analysis of CS (Hope 7B) x CS recombinant lines developed by Law (1966) has allowed us to map the structural gene units controlling the extra band (B1) and CS bands 8 and 12. All are located close together about 6 recombination units from the centromere on the long arm of 7B. This cluster of genes is known as the α-Amy-B2 locus and the Hope allele at this locus is identified as α-Amy-B2b, the 'a' allele being that carried by CS. The homoeologous loci on chromosomes 7A and 7D are known as α-Amy-A2 and α-Amy-D2 respectively.

Similar variants can be found in the α-AMY1 isozymes such as the two extra bands found in <u>Triticum spelta</u> relative to CS (Figure 3). Analysis of the appropriate chromosome substitution series shows the extra bands to be controlled by chromosome 6B. This variant is very common and known as the α-Amy-B1b allele. Close examination of the higher pI bands in Figure 3 reveals additional variation at the α-Amy-A1 and α-Amy-D1 loci.

CONCLUSIONS

These analyses of the wheat α-amylase isozymes indicate that the two groups of enzymes are under independent genetic control. The isozymes with more basic pIs are all encoded by the α-Amy-1 genes on the long arms of the group 6 chromosomes. These loci are linked to the centromere with recombination frequencies of 11.9 ± 2.8% (centromere 6D - α-Amy-D1) and 19.4 ± 3.8% (centromere 6B - α-Amy-B1) (Nishikawa et al. 1981). It is probable that these values represent homoeologous positions and therefore α-Amy-A1 will be found to be similarly located on chromosome 6A. Nishikawa and co-workers also present evidence for a duplication of the α-Amy-B1 locus 20.6 ± 2.2% recombination units distal to the first on 6BL.

Similarly the isozymes with more acidic pIs (αAMY2) are produced by the α-Amy-2 loci on the long arms of the group 7 chromosomes. The α-Amy-B2 locus is linked closely to the centromere of 7B (5.9 ± 5.5%) and α-Amy-A2 and α-Amy-D2 can be expected to be homoeologously positioned on 7A and 7D respectively.

It is of interest to note that the group of enzymes classified as GII with pIs intermediate between the two main bodies of malt and green isozymes by Marchylo, LaCroix and Kruger (1980) are almost certainly α-AMY2 (eg. bands 16 and 17 in Figure 1) and encoded by the same genes as their GI group.

Allelic variants have been found at most of these loci and are described by Nishikawa et al. (1981) or Gale et al. (1982). These variants can be complex involving, as in the case of α-Amy-B2b, the addition of one new isozyme and the absence of two from the CS phenotype. The individual α-Amy-1 and α-Amy-2 loci probably comprise tightly linked complexes of 'genes' coding for several individual isozymes, and may be the products of successive evolutionary duplication and mutation.

Both allelic and homoeologous variation may be exploited in the identification of chromosome regions in chromosome assays for agronomic characters, the checking of intervarietal substitution lines and the identification of alien chromosomes that have been either added to, or substituted for, the hexaploid wheat chromosome complement.

Finally it is probably of value to summarise the various terms used for these enzymes by different workers.

α AMY1	α AMY2	
malt, II	green, I	Daussant, Mayer and Renard - 1981
Group 1	Group 2	Sargeant - 1980
GIII	GI and GII	Marchylo, LaCroix and Kruger - 1980.

REFERENCES

Daussant, J., Mayer, C., and Renard, H.A. 1980. Immunochemistry of cereal α-amylases in studies related to seed maturation and germination. Cereal Res. Commun. 8:49–60.

Gale, M.D., Law, C.N., Chojecki, A.J., and Kempton, R.A. 1982. Genetic control of α-amylase production in wheat. Theoretical and Applied Genetics (in press).

Law, C.N. 1966. The location of genetic factors affecting a quantitative character in wheat. Genetics 53:487–498.

Marchylo, B.A., LaCroix, L.J., and Kruger, J.E. 1980. Synthesis of α-amylase in specific tissues of the immature wheat kernel. Cereal Res. Commun. 8:61–68.

Nishikawa, K., and Nobuhara, M. 1971. Genetic studies of α-amylase isozymes in wheat. I. Location of genes and variation in tetra- and hexaploid wheat. Jpn. J. Genet. 46:345–358.

Nishikawa, K., Futura, Y., Hina, Y., and Yamada, T. 1981. Genetic studies of α-amylase isozymes in wheat. IV. Genetic analysis in hexaploid wheat. Jpn. J. Genet. 56:385–445.

Sargeant, J.G. 1980. α-Amylase isoenzymes and starch degradation. Cereal Res. Commun. 8:77–86.

Differences in Falling Number at Constant Alpha-Amylase Activity

Kåre Ringlund, *Agricultural University of Norway*

ABSTRACT

Differences in falling number are due to differences in alpha-amylase activity and in starch value. These two factors were separated by determining both falling number and alpha-amylase activity, and analyzing falling number in an analysis of covariance with alpha-amylase activity as the covariate. Both measures were transformed to logarithms to obtain a linear relationship between them. Differences in starch value were found between varieties and locations, and between different levels of sprouting damage. On the basis of the results of these investigations, the choice of methods for evaluating sprouting damage is discussed.

INTRODUCTION

Differences in falling number between samples of grain which do not differ in sprouting damage have been reported earlier. Olered (1967) showed that some of this variation in falling number was due to a alpha-amylase which is normally inactivated during ripening, but which might be reactivated under certain climatic conditions. Differences in starch value, i.e. differences in falling number at the same alpha-amylase activity, have been reported by Ringlund (1965) and Olered (1967). Grain with a high starch value will have an acceptable baking quality at a higher level of sprouting damage than grain with a low starch value. The purpose of the investigations in this report, was to evaluate the importance of environmental and genetic differences in starch value.

MATERIAL AND METHODS

This paper is based on two sets of data. Part one comprises 8 samples of spring wheat grown in South-Eastern Norway in 1980. Six of the samples were from 3 breeding lines grown at two locations, and 2 samples were from sister lines grown in a preliminary yield trial.

Fifty percent of the grain of each sample was sprouted by adding water to 35 percent moisture content. The samples were left in a sealed glass jar for 36 hours, and then air dried to 10 percent moisture at 40 C. Using the diastatic activity (Dt)

Dt=6000/(falling number - 50),

as a mixing value, each sample was divided into 10 subsamples (blends) with a precalculated falling number as shown in Table 1.

TABLE 1.
Construction of 10 subsamples with different falling numbers (Fn) from grain samples no.1.

Blend			Wanted		Percent in mixture	
	Fn	Dt	Fn	Dt	original	sprouted
1	350	20			100	0
2			330	21	99	1
3			310	23	90	10
4			290	25	83	17
5			270	27	76	24
6			250	30	67	33
7			230	33	57	43
8			210	38	40	60
9			190	43	24	76
10	170	50			0	100

In order to obtain the wanted falling number, the differences between 350 and 170 were divided into 9 equal increments. This increment was added consecutively to the lowest value. The diastic activities were calculated from the wanted falling numbers. In order to obtain a sample with for example Dt=25, x percent of grain with Dt=20 and (100-x) percent of grain with Dt=50 had to be mixed.

$$20x + (100-x) 50 = 100 25, x=83$$

The blends were ground on a falling number mill. Falling numbers and alpha-amylase activities were determined for each blend.

Part two is based on data from two series of yield trials in spring wheat, grown in 1980 and in 1981. Each series was grown at 6 locations and had 25 varieties or breeding lines.

The falling number was determined by the standard method using 7 grams of ground wheat at 15 percent moisture base, and 25 ml of water. The alpha-amylase activity was determined by a colorimetric method described by Olered (1967).

SEPARATING THE EFFECT OF STARCH VALUE AND THE EFFECT OF ALPHA-AMYLASE ACTIVITY ON FALLING NUMBER.

In previous publications it has been shown that the relationship between falling number (Fn) and alpha-amylase activity $(K(\alpha))$ is approximately linear when both measures are transformed to logaritms (Ringlund (1965), and Olered (1967)). Differences in starch value can be seen in a diagram of ln (Fn) and ln $(K(\alpha))$ as shown in Figure 1.

FIGURE 1.
Differences in starch value.

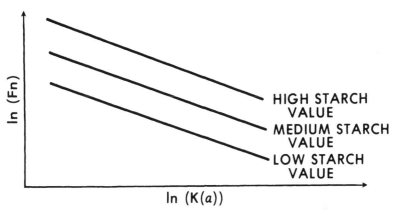

Differences in starch value can be tested by testing differences in levels between regression lines, or by an analysis of covariance for ln (Fn) using ln $(K(\alpha))$ as the covariate. Differences in ln (Fn) between sample means after covariance corrections, are due to differences in the starch values.

RESULTS

The results will be presented separately for part one and part two.

Part one

The falling numbers of samples 1-8 and blends 1-10 are presented in Table 2. For sample No 8 the sprouting treatment was not severe enough to give the intended reduction in falling number. For the other samples a satisfactory range in falling numbers with a corresponding range in alpha-amylase activity was obtained (Table 3).

TABLE 2.
Falling numbers of 8 grain samples with 10 different levels of sprouting.

Blend No	\	\	\	Grain sample No	\	\	\	\	x̄
	1	2	3	4	5	6	7	8	
1	389	411	421	262	312	372	334	262	345
2	373	363	399	248	302	334	313	247	322
3	356	305	396	242	283	308	288	251	304
4	310	253	365	224	265	282	275	238	277
5	269	235	357	220	255	262	241	241	260
6	264	229	331	206	246	226	225	237	246
7	253	200	310	195	212	195	204	239	226
8	214	177	305	181	193	178	184	236	209
9	193	146	273	165	168	139	167	228	185
10	170	109	256	137	140	103	134	232	160
x̄	279	243	341	208	238	240	237	241	253

TABLE 3.
Alpha-amylase activity, $K(\alpha)$, of 8 grain samples at 11 different levels of sprouting.

Blend No	\	\	\	Grain samples No	\	\	\	\	x̄
	1	2	3	4	5	6	7	8	
1	1.30	0.80	0.51	1.48	1.89	0.80	0.99	1.74	1.19
2	1.43	0.78	0.48	1.46	2.61	1.34	1.54	2.33	1.50
3	1.46	1.54	0.84	1.65	2.68	1.68	1.50	2.11	1.68
4	2.76	1.68	0.91	3.00	2.92	2.08	1.67	2.22	2.16
5	2.78	1.95	0.97	2.92	3.34	2.66	1.98	2.39	2.37
6	2.85	3.21	1.00	3.61	3.93	3.43	2.72	2.62	2.92
7	4.84	3.70	1.43	3.26	4.94	4.30	3.31	2.41	3.52
8	5.35	5.02	1.54	4.15	6.69	6.51	4.18	2.14	4.45
9	4.84	7.30	1.76	4.39	7.53	9.87	6.00	2.16	5.48
10	6.70	10.03	2.06	7.48	13.02	15.87	7.13	2.62	8.11
x̄	3.43	3.60	1.15	3.34	4.96	4.85	3.10	2.27	3.34

The relationship between falling number and alpha-amylase activity on the means of the 10 different levels of sprouting damage is shown in Figure 2. The simple linear correlation coefficient is 0.9939.

The effect of germination on starch value. Differences in starch value as defined above must be due to different swelling capacity of the starch of different samples. During germination the starch is degraded by the attack of alpha-amylase. Sprouted grain, therefore, should be expected to have lower starch value than unsprouted grain. This effect can be tested in this material by an analysis of covariance with blends as treatments and samples as replications. The mean falling number and starch values for the 10 blends are given in Table 4. Only blends 9 and 10, show a decrease in starch value, but the decrease is statistically significant. This result is in good agreement with earlier findings that native starch has good resistance to alpha-amylase.

FIGURE 2.
Relationship between ln (Fn) and ln (K(α)).
Means of 10 different sprouting levels.

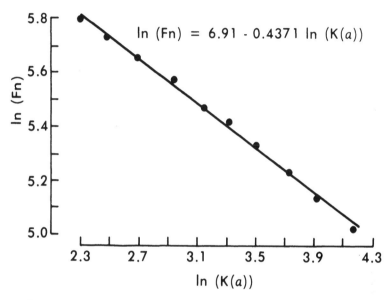

$$\ln (Fn) = 6.91 - 0.4371 \ln (K(a))$$

TABLE 4.
Falling number (Fn) and starch value (Sv)of wheat with different degree of sprouting damage. Retransformed values.

Blend No	1	2	3	4	5	6	7	8	9	10
Fn	340	317	299	273	257	245	224	204	181	153
Sv	247	247	252	250	245	250	247	245	228	221

The difference in starch value between different sprouting levels will influence the comparison of starch values of a nonsprouted sample and a sprouted sample. In this material, where all samples were purposely tested at different sprouting levels, the effect of sprouting will influence the slope of the regression lines for ln (Fn) and ln (K(α)), but not the levels of the regression lines.

Differences between grain samples. Regression lines for 7 of the grain samples are shown in Figure 3. For sample No. 8 the regression coefficient was not significant. There are highly significant differences in levels between the 7 regression lines, but the differences in slopes are not significant.

FIGURE 3.
Linear regression function for different grain samples.

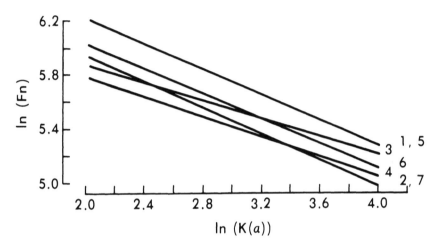

Since samples 1-6 were factorial combinations of 3 varieties and 2 locations, the differences between these samples can be split into a varietal effect, a location effect and an effect of variety x location interaction. The appropriate data are presented in Table 5.

TABLE 5.
The logarithm of starch values, ln (S.V.), and retransformed starch values for 3 varieties at 2 locations.

Variety	Location 1	Location 2	x	Starch value
1	5.650	5.425	5.538	254
2	5.477	5.375	5.426	227
3	5.629	5.509	5.569	262
$\bar{x}$	5.585	5.436	5.511	
Starch value	266	230		247

The variety x location interaction is significant.

Part two

The differences in starch value between the 12 nurseries are highly significant. Splitting the effect of nurseries into locations, years and a year x location interaction shows a significant interaction. The difference between the 2 years is significant with the interaction as error term.

Locations 1, 4 and 5 are in the Oslo-area, whereas locations 2,3 and 6 are 100-120 km further north. The difference in starch values between the two areas (340-308) is significant (Table 6).

The 7 varieties which were in all 12 nurseries, have significantly different starch values (Table 6).

TABLE 6.
Starch values of 7 varieties and 6 locations.

Variety	Starch value	Location	Starch value
1 Reno	296	1	337
2 Runar	330	2	327
3 T638-198	334	3	299
4 MØ 75-711	334	4	344
5 T8026	317	5	344
6 T8033	299	6	305
7 T8035	324		

The earliest maturing varieties have the highest starch values.

DISCUSSION

Differences in starch value have much less drastic influence on falling number than differences in alpha-amylase activity. Nevertheless, the differences found in starch quality contribute significantly to the total quality of the evaluated samples.

The results reported are important for the choice of methods for evaluating sprouting damage. If the purpose is to evaluate a sample for its swelling capacity, the falling number should be used. If the main interest is the diastatic activity of a sample, the determination of alpha-amylase activity on a standard starch solution is appropriate.

REFERENCES

Olered, R., and Jønson, G. 1964. The development of α-amylase activity in ripening wheat. Arkiv. Kemi 22(15):175-183.
Olered, R.]967. Development of α-amylase and falling number in
wheat and rye during ripening. Vaxtodling No. 23:1-106.
Ringlund, K. 1965. Stivelseskvalitet og proteinkvalitet hos nors-
kavlet kveite. Meld. Nor. Lundbrukshoegsk 44(23):1-36.
Ringlund, K. 1980. Starch quality in wheat and barley at different
maturity stages in relation to seed dormancy. Cereal Res.
Commun. 8:193-197.

The Role of the Amylograph in Assessing Sprout Damage

Peter Meredith, *Wheat Research Institute, Christchurch, New Zealand*

INTRODUCTION

The amylograph is undoubtedly the most sensitive test available for assessing sprout damage in cereals. Comments will be confined to assessment of sprout damage in wheat. The AACC Official Methods describe the amylograph as "used primarily to determine the effect of alpha-amylase on viscosity of flour as a function of temperature" (AACC 1982).

The first point to make is that although the amylograph is very sensitive and when properly used is very precise the results from it do not appear in any units that may be directly related to the degree of sprout damage, to user qualities, or even to activity of alpha-amylase. The reasons for this will be examined.

The next point is that there is a marked difference between field sprouting and sprouting simulated in some way by the effect of treatments to induce germination. The belief is simply that there are fewer kinds and lesser quantities of enzymes produced in field sprouting situations than are produced by laboratory germination, malting or by moist sand or mist chamber type of experimental assessment. However, this evidence is circumstantial and will not be discussed in this paper. The differences between field sprouting and artificial conditions are not very apparent when starch is used as a substrate, whether it is starch in the amylograph or in some modified form as an alpha-amylase assay. This is contrasted to the situation when using thermally pasted flour as substrate as far more than just starch is present, i.e., proteins and gums. The importance of gums is well known to those involved in rye-using industries.

The importance of proteins in the pasting situation does not seem to be so widely appreciated. In the amylograph, flour protein causes an appreciable part of the hot paste viscosity. This is readily seen when wheat flour and starch prepared from that wheat flour are compared in an enzyme-free state (Fig. 1). These curves are nearly identical indicating that components of flour other than starch contribute with starch to produce peak pasting strength.

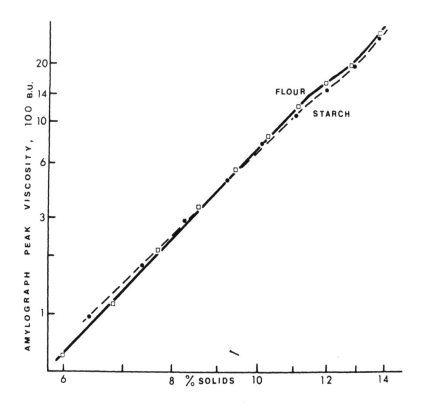

Fig. 1 Relationship of flour and starch solids with amylograph peak viscosity.

Since starch is no more than 80% of flour solids composition, it is evident that starch alone cannot possibly account for the pasting strength of flour. Additional experiments show that proteins of flour possess pasting strengths similar to that of starch. Protein is an important variable (Moss 1967).

Particularly when malt additions are used, but also in other germination procedures, there seems to be an appreciable protease type of activity that thins the pasting ability of the protein component in the assay. However, field sprouting, at least under our New Zealand conditions, seems to have minimal effect on protein and gums; the assay is more affected by amylase activity.

When using flour in the amylograph assay, four variables must be contended with in addition to the presence of various enzymes:
- inherent pasting ability of the starch granules
- susceptibility of undamaged granules to attack by amylase
- protein and gum content
- damaged starch content.

Is it any wonder that since the assay system is completely arbitrary that it is not possible to get sensible units from the assay.

So, for the present experiment the assay is restricted to a single sample of flour, a single amylograph, and a single sample of field-sprouted wheat, in this case from Kansas in 1979. A brief treatment with dilute acid at pH 2.5 can irreversibly inactivate amylases in flour. Although this process was rediscovered (Meredith 1970) and used in the laboratory and industrially, this would be an appropriate time to acknowledge Sandstedt et al. (1937) for discovery of this technique. This acid-treatment was included in a paper having quite other intent and seems to have remained unnoticed.

The amylase-inactivation method enables an examination of the inherent pasting ability of flour components. Active enzyme can be added into the system, thus avoiding many of the caveats that may be levelled against the addition of prepared enzyme systems to arbitrarily synthesised pasting systems.

The relationship between concentration of flour used in an amylograph and peak height observed on pasting is linear when plotted on a log-log basis, within limits, as shown in Figure 1. This linear relationship is quite fortuitous but very useful. At either high concentrations or high viscosities, there is a departure from linearity to another line of similar slope but offset. At low concentrations or low viscosities, a curvature of the relationship may become apparent on careful examination of the lower end of the scale. These relationships have been observed for both New Zealand and U.S. flours.

Thus, by inactivating a proportion of the flour and keeping the remainder in an enzyme active state, a family of curves can be produced showing the effect of enzyme of field sprouting on the sprout-damaged material itself under conditions of pasting that resemble some of the conditions of industrial usage (Fig. 2).

The log-log relationship retains the same slope whatever addition is made. The effect of enzyme addition can be quantified by the degree of offset of the lines of association.

The second observation is that doubling of the amount of active addition gives equal amounts of shift of the log-log curve. Thus, for a given system, two amylograph determinations, one inactivated and one in the fully or partly active state, should enable a calculation of a meaningful unit of enzyme action. That would be fairly easy, but there is a problem.

The problem is that not all flours are equally susceptible to attack by the same enzyme system. The investigation is now concerned with the solution to this problem.

122

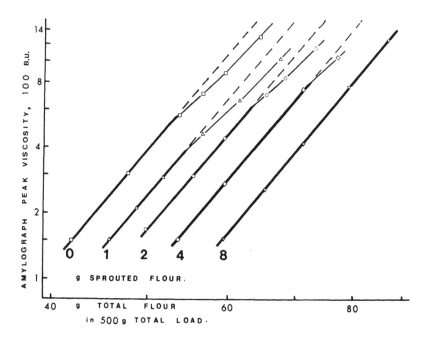

Fig. 2   The effect on amylograph peak viscosity of adding field-sprouted wheat flour to varying proportions of the same flour that has been acid-treated.

The obvious variables are:

- proteins and gums being possible contributors to pasting and being attacked by some enzyme systems but not attacked by others;
- the inherent variation of pasting ability of starches, which is not a major problem since all give approximately parallel log-log curves;
- the varying susceptibility of starch of a flour to attack, both because of a variation of inherent susceptibility of the intact starch to amylase attack and because of varying degrees of mechanical damage from milling that will always give a high degree of susceptibility to attack in damaged granules.

But the nature of this family of curves reveals more yet. The parallel lines are given by additions of equal amount of amylase to the total mixture in the amylograph pot, not by equal proportions of amylase to flour; that only gives curious curves. This suggests that it is the concentration of enzyme in the liquid that is important, rather than the proportion of enzyme to substrate.

This discussion has been deliberately restricted to hot pasting peak viscosity. This is of course only one aspect of the use of flour and starch as a thickener. The properties of the cooled gel are of equal importance and not necessarily related to the hot paste properties, but these will not be discussed here.

To summarise, there exists a rough and ready expresion of enzyme activity as measured by the amylograph. These are the Amylograph Units that were published a good many years ago. Looking at the present situation with a great deal more understanding, we have a complex situation and a very sensitive method that is, moreover, close to some of the real life usage situations. Precise units measuring enzyme activities and the degree of sprout damage are possible but are not available at present. Consequently, to determine the degree of sprout damage for bread baking purposes, there is only one certain method. That is to mill flour and bake bread under standard conditions and to assess the stickiness and loss of resilience of the crumb.

In our experience, the only laboratory test giving results that parallel the bake test under a wide variety of field sprouting conditions is the old Hagberg Penetrometer Test, where a conical penetrometer is dropped into a cold porridge of flour and water after cooking and cooling. In other words, for breadmaking purposes, it is the properties of the cooled gel, as mentioned above, that may be of far greater importance than the hot paste viscosity. That may be one reason for the occasional failure of the Hagberg Falling Number method to agree with commercial experience.

Aims and objectives must be clear. Is the objective to determine the degree of change that may have taken place in the grain as the result of inclement weather conditions or is the objective to measure the suitability of the grain for some particular end-use? The amylograph is not a suitable technique for determining the changes in grain, unless it is used in a very precise manner to measure all the possible changes as separate factors. But when measuring suitability for a hot pasting or cooled gel end-use, then it is perhaps much more suitable than any other method available. The limitation is that the amylograph test is time consuming and, therefore, not good as a rapid screening test. Thus, the following tests are recommended:

- FALLING NUMBER is highly suitable as a rapid screening test of hot pasting ability;
- the PENETROMETER test is very suitable to test cold gelling property;
- any of the newer rapid ALPHA-AMYLASE assays are suitable as tests of grain damage;
- the AMYLOGRAPH, when used as a single test, can be very misleading.

The AMYLOGRAPH used as an INVESTIGATIVE PROCEDURE can give the best information of all, enabling an assessment of several independently variable factors:

- inherent hot pasting strength of starch;
- inherent cold gelling strength of starch;
- contribution of protein and gums to pasting and gelling;
- contribution of mechanically damaged starch;
- susceptibilities of undamaged and damaged granules to enzyme attack;
- effects of amylases and of other enzymes, present as a result of field sprouting, on all of the above variables.

It is a very complex technological situation. No simple test can ever be more than a rough guide.

REFERENCES

American Association of Cereal Chemists, Approved Methods. 1982. Vol 1. AACC Method 22-10.

Meredith, P. 1970. Inactivation of cereal alpha-amylase by brief acidification: the pasting strength of wheat flour. Cereal Chem. 47:492-500.

Moss, H.J. 1967. Flour paste viscosities of some Australian wheats. J. Sci. Fd. Agr. 18:610-612.

Sandstedt, R.M., Blish, M.J., Meecham, D.K., and Bode, C.E. 1937. Identification and measurement of factors governing diastasis in wheat flour. Cereal Chem. 14:17-34.

Some Experiences with Monitoring Alpha-Amylase Levels in, Canadian Wheat

J. E. Kruger and K. H. Tipples,
Grain Research Laboratory, Canadian Grain Commission, Winnipeg, Manitoba, Canada

SUMMARY

In addition to the Amylograph and the Falling Number Methods, our Laboratory has recently implemented a specific α-amylase method into the monitoring program for Canadian cereals. This method has provided new information on: i) the effectiveness of visual sprout assessment for 1 CWRS wheats; ii) levels of α-amylase in damaged kernels representing different degrading factors; iii) the influence of severity of sprouting upon distribution of the enzyme in mill fractions; and iv) assessment of sprout-resistant cultivars.

INTRODUCTION

For a number of years the Grain Research Laboratory has monitored levels of the enzyme α-amylase in top grades of Canadian wheat in cargoes and carlot unloads at our terminal elevators, using the Hagberg Falling Number Method, carried out on ground wheat, and the Brabender Amylograph Method, carried out on the flour. Such methods, being autolytic in nature, may be influenced by factors such as starch damage, nature of the starch, fiber, etc. To augment such tests, therefore, we have in the last few years implemented methods more specific for α-amylase. The method initially used was an automated fluorometric procedure using a Technicon AutoAnalyzer, with β-limit dextrin anthranilate as substrate (Marchylo and Kruger 1978). Recently we have switched over to using a nephelometric method, using the Perkin Elmer Model 191 Grain Amylase Analyzer (GAA). This has proven to be simpler to use and less expensive, although sample throughput is slower. In this paper we would like to discuss some of our experiences with this new method and illustrate some of the implications for our monitoring program for α-amylase levels in wheat.

Paper No. 491, Canadian Grain Commission, Grain Research Laboratory, 1404-303 Main Street, Winnipeg, Manitoba R3C 3G8.

MATERIALS AND METHODS

Assay of α-Amylase. The procedure used is based on that described by Campbell (1978) and employing the Model 191 Grain Amylase
Analyzer. Because the levels of α-amylase in Canadian cultivars
harvested under normal weather conditions are very low, it was
necessary to modify the standard Model 191 GAA procedure in order to
make it more sensitive (Kruger and Tipples 1981). This was accomplished by: a) increasing the sensitivity on the machine to its
maximum; b) reducing the substrate concentration to 0.5%; c)
attaching a recorder in order to more precisely follow the decreases in turbidity. This modified procedure could adequately
detect α-amylase in flours with amylograph viscosities up to 925 BU
and wheats with falling numbers as high as 470 sec.

The substrate, β-limit dextrin, is prepared in our laboratory
by treating waxy-maize starch with β-amylase followed by concentration, dialysis and freeze-drying. Because of its importance, the
nature of this substrate has been examined further. One of our
findings has been that although a very consistent substrate can be
prepared from the same batch of waxy-maize starch, differences have
been found in the response of β-limit dextrin prepared from two
separate batches of waxy-maize starch. This indicates that there are
differences in the molecular structure of different sources of waxy-
maize starch that should be taken into account in preparation of
β-limit dextrin. Research has also been carried out on the mechanism of breakdown of this substrate by wheat and fungal α-amylases
using high performance aqueous gel permeation chromatography (Kruger
and Marchylo 1982). A spherogel TSK 3000 SW column with an exclu-
sion limit of 1.5×10^5 daltons has been used to analyze the molecular weight distribution of products formed. The breakdown of β-limit
dextrin proceeds in stages. Initially a very rapid breakdown of
substrate occurs, which results in the formation of high molecular
weight dextrin products, larger than 1.5×10^5 daltons and with
little production of low molecular weight dextrins and sugars. It
is during this phase of the reaction that the turbidity of the substrate is lost and this is the basis of the nephelometric method.
In the next stage of the reaction intermediate molecular weight products are formed which slowly decrease in size concurrent with a
progressive build-up of low molecular weight sugars and oligosac-
charides. This phase of the reaction is the important one for
methods based upon production of reducing sugars and explains why
such methods would be less sensitive than the nephelometric method.
Malted HRS and durum wheat α-amylases appeared to have identical
action patterns, which in turn were different from that catalyzed by
fungal α-amylase.

RESULTS AND DISCUSSION

Effectiveness of visual grading for 1 CWRS wheats. To evaluate
just how effective visual assessment of sprouting and weathering has
been in maintaining low levels of α-amylase in our top grades of

wheat, 138 carlots graded as 1 CWRS on unload at terminal elevators
were sampled and tested for α-amylase activity. The carlots represented 1979 crop wheat from primary delivery points across Western
Canada (Fig. 1). All contained less than 0.5% sprout damage, the
maximum allowable at the country level for the grade No. 1 CWRS.

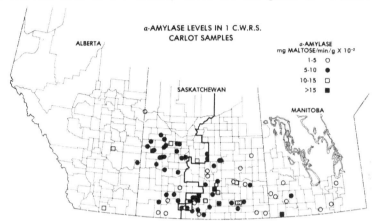

Figure 1. α-amylase levels in 1 CWRS carlot samples from delivery
points across Western Canada.

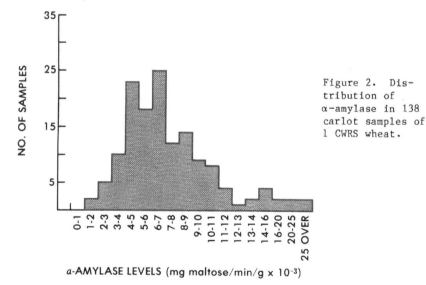

Figure 2. Dis-
tribution of
α-amylase in 138
carlot samples of
1 CWRS wheat.

One hundred and thirteen (113) of the samples had α-amylase
levels less than 10 mg maltose/min/mg x 10^{-3}, which corresponded to

a falling number of 340 seconds or higher. Of these, a large proportion had activities of 4-7 mg maltose/min/mg x 10^{-3} (Fig. 2). Nineteen of the samples had an activity greater than 10 but less than 15 mg maltose/min/mg corresponding to a falling number of 320 seconds or higher. Six samples had higher levels of activity and on further examination we found that two of them had been misgraded and were, in reality, No. 2 CWRS. The other four samples came from a region representing the major area for growing the sawfly-resistant cultivar, Chester. This cultivar had a higher endogenous level of α-amylase than other cultivars without showing evidence of sprouting. Although grown in only small quantities, this cultivar is now being phased out of commercial production. The new cultivar Leader, represents the first of a new series of sawfly-resistant lines having low alpha-amylase activity and high sprouting resistance. The conclusion thus reached from this study was that our visual grading system was indeed effective in maintaining the amount of enzyme in our top grade of wheat at a very low level. Rye on the other hand, is more difficult to assess visually and the correlation coefficient between α-amylase and % sprouted kernels for 82 samples was only 0.4 (Kruger and Tipples 1982). In rye, factors such as the presence of degermed or skinned kernels make visual assessment of sprouting, and hence α-amylase activity, more difficult.

α-Amylase levels in damaged wheat kernels representing different degrading factors. Although elevated levels of α-amylase are normally associated with presence of sprouted kernels, it is of equal concern when setting visual grading standards to know whether

TABLE 1

ALPHA-AMYLASE (MG MALTOSE/MIN/G X 10^{-3}) IN DAMAGED WHEAT KERNELS

STATION	3 CWRS WHEAT					3 CU WHEAT				
	KIPLING	ROWANVILLE	MAKINAK	PORTAGE		VIKING	PELLY	KAMSACK	FT. SASK.	
SOUND	1.4	1.6	2.4	2.5	2.3	2.8	3.5	1.5	3.3	5.3
DEGRADING FACTOR										
NON-VITREOUS	—	—	—	2.4	—	—	1.7	—	—	—
BLEACHED	—	—	2.1	—	—	4.2	—	—	—	—
SEVERE BLEACHED	—	—	9.4	—	—	—	2.3	1.8	—	—
FROST	—	—	—	—	—	—	—	2.9	—	—
HEAVY FROST	1.7	—	—	—	—	—	—	—	—	—
SHRUNKEN, IMMATURE, GREEN	25.0	—	7.3	11.8	—	100.0	—	38.5	—	—
SHRUNKEN, IMMATURE	—	6.9	—	—	—	—	—	—	—	402.0
MILDEW (SOUND)	2.5	2.5	1.4	—	—	5.0	10.7	5.4	22.8	—
MILDEW (SUSPECTED GERM.)	2.3	—	2.8	—	201.0	—	—	—	—	21.8
BLACKPOINT	—	—	—	55.0	640.0	—	—	—	147.0	—
SPROUTED	—	—	—	—	—	—	8820.0	—	9550.0	9810.0

high α-amylase activity may be present in other types of damaged
kernel. To this end, samples of wheat from different country elevators representing 3 CWRS and 3 CU wheats were hand-separated to provide kernels representing a number of degrading factors present in
wheat. The original sound wheat, along with the hand-picked damaged
kernels were analyzed for α-amylase using the Model 191 Grain Amylase Analyzer. As shown in Table I, non-vitreous, bleached, and
lightly frost-damaged kernels contained normal, low levels of
α-amylase. Shrunken, immature green kernels had higher levels of
activity. Mildew by itself had no noticeable effect, unless it was
associated with visible signs of germination. It was interesting
that kernels showing blackpoint had relatively elevated levels of
activity although much lower than those found in germinated kernels.
Further work is planned with a wider range of samples to ascertain
variability in α-amylase activity that can be expected to be associated with damaged kernels of different origin.

Influence of severity of sprouting upon the distribution of
α-amylase in mill fractions. For many years we have noticed variations in relationships between wheat falling number and flour amylograph viscosity. For example, two wheat samples may have similar
falling number values but their flours may have quite different amylograph viscosities. By carrying out α-amylase determinations on
both the wheats and the flours milled from them, we have found that
there can be substantial variations in the amount of α-amylase that
can be removed in the milling process. To put it another way, the
proportion of total wheat amylase activity present in the endosperm
can vary widely. We have examined a number of possible reasons for
this variation but so far have come up with no firm conclusion as to
the cause (Kruger and Tipples 1980). A recent study carried out on
the distribution of α-amylase in millstreams from wheats sprouted to
different extents suggests, however, that severity of sprouting
could be a factor in this variability (Kruger 1981). Thus, it is
known that α-amylase is synthesized initially in the scutellum and
aleurone layers. If the wheat is exposed to conditions favoring
germination for a greater length of time, the α-amylase begins to
move progressively into the endosperm. When this happens it is
removed less effectively with the branny layers when the wheat is
milled. This hypothesis was tested on our pilot mill by comparing a
sample of sound wheat, to which had been added 1.8% of 44 hr germinated wheat, with a sample formed by adding 0.3% of more severely
sprouted (96 hr germinated) wheat to the sound wheat such that the
final total wheat α-amylase activities were very similar. It was
found that the distribution of α-amylase in the millstreams of the
two wheats was quite different. The sample containing a smaller
amount of more severely sprouted wheat had more enzyme in the first
break flour on a per unit weight basis, more in the sizings, but
less in the later middlings than the lightly sprouted wheat. This
indicated that the enzyme had penetrated more into the endosperm in
the severely-sprouted sample. To see if this pattern also occurred
with wheats that undergo field sprouting (as contrasted to
laboratory-germinated samples), we compared a sample of 1 CWRS wheat
from the 1978 crop with a sample of 2 CWRS wheat from the 1979 crop.

The level of enzyme in the original wheats was similar. As 1978 was a wet harvest year compared to 1979 it would be expected that the severity of sprouting would be greater and the very small number of sprouted kernels that find their way into the top grades would have proportionately more α-amylase in the endosperm. The distribution of α-amylase in the mill fractions indicated more enzyme in the break streams from the 1978 wheat than the 1979 wheats but less in the middlings streams. Thus, samples containing small amounts of field-sprouted kernels gave similar results to those obtained with laboratory-sprouted wheats. This further reinforced the view that severity of sprouting is a factor influencing how much of the total wheat α-amylase level will end up in a milled flour.

Assessment of sprout-resistant cultivars. Following an extremely wet harvest in 1968, Canadian wheat breeders increased their efforts to improve weathering and sprout resistance in spring wheat. The cultivar Columbus, licensed in 1980, represented the first of a new generation of cultivars with improved sprout resistance and lower endogenous α-amylase. With the development of these improved lines it became more difficult to discriminate between cultivars. We have found the modified Model 191 GAA procedure useful in supplementing information provided by the falling number and amylograph methods. It is particularly useful in a dry harvest year when the delineation between cultivars being tested for licensing is not too distinct. By carrying out a specific α-amylase assay we can determine whether an increase in amylograph viscosity is due to a change in starch characteristics or to α-amylase alone. As the acreage seeded to new sprout- and weather-resistant cultivars increases, the α-amylase activity of top grades may be expected to decrease even further and will eliminate any question of supplementing the visual grading with a segregation based on α-amylase activity measurement.

CONCLUSIONS

Because of their world-wide acceptability, the Hagberg Falling Number and Amylograph Methods will continue to be used to monitor α-amylase in Canadian wheat. More laboratories may be expected to begin to report results in terms of α-amylase activity and the research that we are presently carrying out with the Model 191 GAA will hopefully provide an impetus for such a trend.

REFERENCES

Campbell, J.A. 1980. A new method for detection of sprout-damaged wheat using a nephelometric determination of α-amylase activity. Cereal Res. Commun. 8:107-113.

Kruger, J.E. 1981. Severity of sprouting as a factor influencing the distribution of α-amylase in pilot mill streams. Can. J. Plant Sci. 61:817-828.

Kruger, J.E. and Marchylo, B.A. 1982. High performance aqueous gel permeation chromatographic analysis of β-limit dextrin hydrolysis by malted hard red spring wheat, malted durum wheat and fungal (Aspergillus oryzae) α-amylases. Cereal Chem. (in press).

Kruger, J.E. and Tipples, K.H. 1980. Relationships between falling number, amylograph viscosity and α-amylase activity in Canadian wheat. Cereal Res. Commun. 8:97-105.

Kruger, J.E. and Tipples, K.H. 1981. A modified procedure for use of the Perkin Elmer Model 191 Grain Amylase Analyzer in determining low levels of α-amylase in wheats and flours. Cereal Chem. 58:271-274.

Kruger, J.E. and Tipples, K.H. 1982. Comparison of the Hagberg falling number and modified Grain Amylase Analyzer methods for estimating sprout-damage in rye. Can. J. Plant Sci. (in press).

Marchylo, B. and Kruger, J.E. 1978. A sensitive automated method for the determination of α-amylase in wheat flour. Cereal Chem. 55:188-191.

Peroxidases and Their Relationship to Dormancy and Germination in the Wheat Kernel

J. S. Noll, *Agriculture Canada, Research Station, Winnipeg, Manitoba, Canada*

SUMMARY

An association between grain dormancy and peroxidase activity was found within a cultivar and among cultivars with different levels of dormancy. Columbus, a hard red spring wheat with harvest-time sprouting resistance resembled its dormant parent, RL 4137 in peroxidase activity more than its recurrent parent, Neepawa. Both anionic and cationic peroxidases were examined by electrophoresis and activity of one anionic and one cationic isoenzyme was different in dormant and non-dormant grains.

INTRODUCTION

Peroxidases are abundantly distributed in plants, including cereal grains (Sullivan, 1946). Their role in the metabolic processes of cereal grains is not understood, however their broad substrate specificity indicates an extremely diverse function. Peroxidases have been implicated in many physiological processes including cell elongation (McCune, 1961), ion transport, cell wall synthesis, ethylene formation and hormone oxidation (Lamport, 1970). Gaspar et al (1977) recently examined the perioxidases in relation to germination of dormant and non-dormant wheats. In isolated embryos, a positive correlation was observed between high peroxidase activity and dormancy. Cold treatment, which is often used to break dormancy, was found to lower peroxidase activity, but the higher peroxidase activity and dormancy relationship, in the embryo, was maintained. On whole grains no such association was observed between dormancy and peroxidase activity. In this latter study, a fast moving cationic peroxidase isoenzyme was reported to be present in the most dormant variety.

A preliminary study of peroxidase activity in grains of Columbus, a recently licensed hard red spring (HRS) wheat with sprouting resistance (Campbell and Czarnecki, 1981), indicated that dormant grains have higher enzyme activity than non-dormant grains. Consequently, the present investigation was undertaken to examine the peroxidases in dormant and non-dormant grains of Columbus and its parents.

MATERIALS AND METHODS

The HRS wheats selected for the study were Columbus and its parents, Neepawa and line RL 4137. Grains of Neepawa have a moderate level of dormancy at harvest time, whereas Columbus resembles its sprouting resistant parent, RL 4137, in its dormancy characteristic. All samples were field grown in the 1979 and 1981 crop years. Seeds from the 1979 crop were stored at ambient temperatures and were considered to be non-dormant (ND) after two and one-half years of storage. Samples from the 1981 crop were harvested at maturity and immediately placed into sub-zero temperature storage (-20°C) to maintain grain dormancy. Semi-dormant (SD) samples were obtained by removing the dormant (D) samples from cold storage for approximately a month before the germination and enzyme tests.

The chilling and germination treatments were carried out on moistened filter paper in petri dishes, on lots of 50 grains. Chilling was accomplished by placing the seed samples on water moistened filter papers and placing them in a cold room (2-4°C) for 48 hours. Germination tests were conducted on both chilled and non-chilled samples at 20°C and 30°C with distilled water or 10^{-4}M gibberellic acid solution (potassium salt of GA_3, from Calbiochem, Calif.) for 48 hours. Germination counts were recorded, but the germinated grains were not discarded. After the various treatments, all samples were frozen, freeze-dried and stored at -20°C until required for the peroxidase assay or electrophoresis.

Peroxidase extracts for the assay were prepared by grinding 10 grains with a mortar and pestle and extracting the ground samples with 10 ml of 10% sucrose solution in stoppered centrifuge tubes, on a horizontal shaker. After the 30 minute extraction period, the samples were centrifuged (12,000 x g, 10 min) and filtered through a Whatman No. 1 filter paper. The filtered supernatants were assayed for peroxidase activity. For enzyme extraction from germinated samples, it was ensured that the appropriate number of germinated grains were included in the sample to correspond to the germination percentages. Enzyme extracts for electrophoresis were prepared as for the assay, except the ground 10 grains were extracted with only 2 ml of sucrose solution.

Peroxidase activity was assayed according to the method described in the Worthington Enzyme Manual, using O-dianisidine as the hydrogen donor (Anonymous, 1972). One unit of peroxidase activity was defined as that amount of enzyme that consumed 1 u mole of hydrogen peroxide per minute at 20°C. Anionic peroxidases were separated by electrophoresis at pH 8.9 according to Davis (1964), applying 300 ul aliquot of extract to each tube and using bromophenol blue as the electrophoretic marker dye. Cationic peroxidases were separated by the same method except the electrodes were reversed. For the cationic peroxidases, only a 20 ul sample was applied to each tube and crystal violet was used as the marker dye. Peroxidase isoenzymes were detected according to the method of Kobrehel and Feillet (1975), using catechol as the hydrogen donor.

The results of the germination tests and the peroxidase assays are the mean of duplicate analyses.

RESULTS AND DISCUSSION

Percent germination and dormancy levels of the various samples used in this study are presented in Tables 1 and 2. The results of the germination data support earlier findings that sub-zero temperature storage maintains grain dormancy and that high germination temperature can induce dormancy in the wheat grain (Noll and Czarnecki, 1980).

Table 1. Percent germination of grains, of the three wheat cultivars at different levels of dormancy, at 20°C for 48 hours with water, gibberellic acid and after chilling treatment.

Sample		H_2O	GA_3	% Germination (±2) Chilled (48 h, 4°C)
Neepawa	– ND	99	97	–
	– SD	87	–	99
	– D	13	47	95
Columbus	– ND	100	99	–
	– SD	20	–	90
	– D	4	26	35
RL 4137	– ND	99	100	–
	– SD	6	–	92
	– D	0	27	35

Table 2. Percent germination of semi-dormant samples at 30°C, for 48 hours with water, gibberellic acid and after chilling treatment.

Sample		H_2O	GA_3	% Germination (±2) Chilled (48 h, 4°C)
Neepawa	– SD	10	77	92
Columbus	– SD	6	33	77
RL 4137	– SD	0	36	84

Table 3 compares the peroxidase activity of the untreated and chilled grain samples. The non-dormant grains have the lowest peroxidase activity. No correlation exists between dormant and non-dormant cultivars when activities of non-dormant samples were compared. Generally, high peroxidase activity paralleled the higher degree of dormancy within a cultivar. Except for the non-dormant samples, peroxidase levels for the two dormant cultivars,

Columbus and RL 4137, were higher than for Neepawa. Although chilling is often used to break dormancy, peroxidase activity of chilled samples was slightly higher than the activity of the corresponding untreated dormant and semi-dormant samples.

Table 4 presents the peroxidase activity of grains germinated at 20°C. H_2O-germinated Neepawa showed a decrease in peroxidase activity with increasing level of dormancy. For Columbus and RL 4137, the semi-dormant samples had the lowest peroxidase activity and non-dormant samples had the highest peroxidase levels for H_2O-germinated grain. Gibberellic acid and chilling treatments both increased peroxidase activity in comparison to samples germinated with water. Peroxidase activity patterns of chilled-germinated samples were once again different for Neepawa and the two

Table 3. Peroxidase activity of untreated and chilled grains, of the three cultivars at different degrees of dormancy.

Sample		Peroxidase activity, units/kernel (±0.06)	
		Untreated	Chilled (48 h, 4°C)
Neepawa	- ND	0.78	-
	- SD	0.97	1.01
	- D	1.05	1.38
Columbus	- ND	1.01	-
	- SD	1.59	1.39
	- D	1.49	1.54
RL 4137	- ND	0.73	-
	- SD	1.20	1.23
	- D	1.40	1.57

Table 4. Peroxidase activity of grains germinated at 20°C, of the three cultivars at different degrees of dormancy.

Sample		Peroxidase activity, units/kernel (±0.06)		
		H_2O	GA_3	Chilled (48 h, 4°C)
Neepawa	- ND	2.18	3.31	-
	- SD	1.89	-	2.52
	- D	1.47	1.80	3.51
Columbus	- ND	2.69	3.25	-
	- SD	1.40	-	2.70
	- D	1.84	2.15	2.04
RL 4137	- ND	2.31	3.01	-
	- SD	1.04	-	2.41
	- D	1.67	1.92	2.07

Table 5. Peroxidase activity of semi-dormant grains germinated at 30°C, for 48 hours with water, gibberellic acid and after chilling treatment.

Sample	Peroxidase activity, units/kernel (±0.06)		
	H_2O	GA_3	Chilled (48 h, 4°C)
Neepawa - SD	1.30	2.09	3.89
Columbus - SD	1.29	1.85	3.01
RL 4137 - SD	1.16	1.89	2.65

dormant cultivars. The activity of dormant Neepawa was approximately forty percent greater than for the semi-dormant samples. In contrast, the peroxidase activity of the dormant samples of Columbus and RL 4137 were approximately twenty-five percent less than for the semi-dormant samples. Semi-dormant grains germinated at 30°C with water, gibberellic acid and after chilling had generally higher peroxidase activity with lower degree of dormancy among the cultivars; Neepawa had the highest enzyme level in the three treatments (Table 5).

Germinated grains had higher peroxidase levels than non-germinated grains. This result was not unexpected and is in agreement with the work of Kruger and La Berge (1974). The dormancy level of the samples used by Gaspar et al (1977) is comparable to the semi-dormant samples in the present investigation. On whole germinated grains, these latter workers found no correlation between dormancy and peroxidase activity, whereas in this study high activity paralleled low degree of dormancy among the germinated semi-dormant samples. Peroxidase activity patterns of grains with different levels of dormancy, within a cultivar, were different for non-germinated (Table 3) and germinated (Table 4) samples. For example, high peroxidase activity paralleled the higher level of dormancy in untreated Neepawa, whereas for germinated grain this pattern was reversed. Peroxidases appear to play some role in grain dormancy since the cultivar Columbus resembles its dormant parent, RL 4137 in peroxidase levels more than its recurrent parent, Neepawa.

To elucidate some of the differences observed in peroxidase activity among the three HRS cultivars and between dormant and non-dormant grains, the peroxidase isoenzymes were examined by electrophoresis. The anionic peroxidases showed very little activity in dormant and non-dormant grains and only a few of the faint bands were visible on photographic reproduction. Figure 1 illustrates the anionic peroxidase patterns for non-dormant and dormant grains of Neepawa and compares the patterns of H_2O-germinated grains of the three cultivars. The only difference observed in the anionic peroxidases between dormant and non-dormant samples was the higher band intensity of one of the slow moving isoenzymes (illustrated by arrow) in the dormant samples. This pattern was the same for all three cultivars. The anionic iso-

enzyme patterns for germinated grain were similar for the three
cultivars, except the intensity of most bands for Neepawa was
greater than that of the other two cultivars. This difference in
intensity was most noticable with the two fastest moving isoenzymes
(Fig. 1), which have much lower activity in Columbus and RL 4137,
the two dormant cultivars. Many of the anionic isoenzymes in
germinated grain were present in the non-germinated samples, but at
such low concentrations that they could not be reproduced photo-
graphically.

Cationic peroxidases have been used to study the inheritance
(Benito et al, 1980) and chromosome location of genes responsible
for wheat grain peroxidases (Kobrehel and Feillet, 1975). Consequently the cationic isoenzymes were also examined in the assump-
tion that they may reflect the differences in peroxidase activity

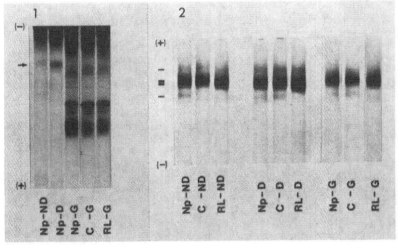

Figure 1. Anionic peroxidase patterns of non-dormant (ND) and
dormant (D) grains of Neepawa and germinated (G) grains of
Neepawa (Np), Columbus (C) and RL 4137 (RL).

Figure 2. Cationic peroxidase patterns of non-dormant, dormant and
germinated grains of Neepawa, Columbus and RL 4137.

between dormant and non-dormant samples. Figure 2 shows the
cationic peroxidases of non-dormant, dormant and germinated grain
of the three cultivars. Four major cationic isoenzymes were
observed. RL 4137 lacks the fastest moving isoenzyme in dormant
and non-dormant grain. However, this isoenzyme is present in
germinated grain of RL 4137. The two intermediate moving cationic
peroxidases have similar mobilities in all samples and appear as a
single band in the photograph. These two isoenzymes appear to
account for most of the activity among the peroxidases and no dis-
tinguishable difference was observed between dormant and non-dormant samples nor among the cultivars. The only difference evident

in the cationic peroxidase patterns between dormant and non-dormant samples was the intensity of the slowest moving isoenzyme band. Band intensity was highest in the dormant samples. In non-dormant samples, the intensity of this slow moving isoenzyme was greatly reduced, whereas in germinated grain it could not be detected. The cationic peroxidase patterns were similar for the three cultivars at the different levels of dormancy, except for the absence of the fastest moving isoenzyme in non-germinated grain of RL 4137. Grain samples exposed to chilling and thus having no dormancy had similar isoenzyme patterns to dormant and non-dormant grain. No difference in peroxidase isoenzyme patterns were observed, that could be associated with dormancy among the cultivars. However, within a cultivar, the activity (band intensity) of one anionic and one cationic peroxidase isoenzyme appeared to be related to grain dormancy. No fast moving cationic isoenzyme was present in the dormant samples as was observed by Gaspar et al (1977).

In summary, an association appears to exist between peroxidase activity and grain dormancy. This was reflected both within a cultivar and among cultivars with different degrees of dormancy. Columbus, the first Canadian HRS wheat with harvest-time sprouting resistance was similar to its dormant parent, RL 4137 in peroxidase levels. The importance of this association between grain dormancy and peroxidase activity cannot be assessed accurately since the native substrates of these enzymes in the grain are not known. However, their broad substrate specificity, especially their possible action on hormones, may indeed be important to dormancy in the wheat kernel.

REFERENCES

Anonymous. 1972. Worthington Enzyme Manual. Worthington Biochemical Corp.; Freehold, N.J.

Benito, C., Perez de la Vega, M. and Salinas, J. 1980. The inheritance of wheat kernel peroxidases. J. Hered. 71: 416-418.

Campbell, A.B. and Czarnecki, E.M. 1981. Columbus hard red spring wheat. Can. J. Pl. Sci. 61: 147-148.

Davis, B.J. 1964. Disc electrophoresis. II Method and application to human serum proteins. Ann. N.Y. Acad. Sci. 121: 404-427.

Gaspar, T., Wyndaele, R., Bouchet, M. and Ceulemans, E. 1977. Peroxidase and -amylase activities in relation to germination of dormant and non-dormant wheat. Physiol. Plant 40: 11-14.

Kobrehel, K. and Feillet, P. 1975. Identification of genomes and chromosomes involved in peroxidase synthesis of wheat seeds. Can. J. Bot. 53: 2336-2344.

Kruger, J.E. and LaBerge, D.E. 1974. Changes in peroxidase activity and peroxidase isozymes of wheat during germination. Cereal Chem. 51: 578-585.

Lamport, D.T.A. 1970. Cell wall metabolism. Ann. Rev. Plant Physiol. 21: 235-270.

McCune, D.C. 1961. Multiple peroxidases in corn. Ann. N.Y. Acad. Sci. 94: 724-730.

Noll, J.S. and Czarnecki, E. 1980. Methods of extending the test-
    ing period for harvest-time dormancy in wheat. Cereal Res.
    Comm. 8: 233-238.
Sullivan, B. 1946. Oxidizing enzyme systems of wheat and flour,
    p. 215-230. In: J.A. Anderson (ed.) Enzymes and Their Role in
    Wheat Technology. Interscience Publishers, Inc. N.Y. 371 p.

Nephelometric Determination of Added Alpha-Amylase in Cereal-Based Products

E. H. Asp[1], *University of Minnesota, St. Paul, Minnesota, U.S.A.*

P. M. Ranum, *Pennwalt Corp., Plymouth, Minnesota, U.S.A.*

L. Tooker Midness[2], *University of Minnesota, St. Paul, Minnesota, U.S.A.*

SUMMARY

The nephelometric measurement of α-amylase activity using the Perkin-Elmer Model 191 Grain Amylase Analyzer is a relatively new and rapid method for measuring the activity of this enzyme in cereal grains and their products. Two factorial experiments were designed to study factors related to new procedures developed to extend use of the method to yeast doughs. Factors that did not seem to affect enzyme activity were sample weight, extraction time, and enzyme stability during 24 h. Factors that did affect measurement of enzyme activity were substrate preparation and mode of the instrument. A procedure was developed to study changes in the α-amylase activity in fermenting and baking yeast dough.

INTRODUCTION

The nephelometric method for measuring α-amylase activity in cereal grains and their products is easier and faster than others presently being used. It detects low levels of activity, and it measures activity of heat-sensitive fungal α-amylase because heat is not used in sample preparation. Major uses of the method by cereal chemists include measuring α-amylase activity in sound and sprouted wheats and their flours, flours supplemented with α-amylase concentrates, and commercial α-amylase concentrates.

In the nephelometric method, the change in light scattering of a β-limit dextrin substrate suspension is measured after hydrolysis by added α-amylase during a timed interval. The instrument designed for use with cereals and cereal products is the Perkin-Elmer Model 191 Grain Amylase Analyzer (GAA). The method was adapted from the clinical procedure described by Zinterhofer et al. (1973). Its use for cereals was reported by Prasad et al. (1979) to measure α-amylase activity for sound and sprouted wheat flours and their blends, and by Kruger, Ranum, and MacGregor (1979) for wheat, flour, barley, and barley malt. The amylopectin substrate used in these early studies was not specific for α-amylase in the presence of β-amylase. Kruger, Ranum and MacGregor (1979) recommended that a β-limit dextrin substrate be used to make the method specific for α-amylase, and that the sensitivity of the GAA be increased. These modifications were made in the nephelometric method described by Campbell (1980a), (1980b).

Studies evaluating the method or modifying it for specific uses

---

[1] Assistant Professor.   [2] Research Assistant.

have since been reported. O'Connell, Rubenthaler, and Murbach (1980) found the 2-minute automatic mode of the GAA more sensitive than the 1-minute mode, and non-linearity of measurements in the 1-mode above 720 GAA units. They used the manual mode to study reaction velocity and enzyme concentration. A procedure was reported by Kruger and Tipples (1981) to increase the sensitivity of the GAA that gave results comparable to those from the automatic fluorometric method for wheat and flour with low to moderate α-amylase levels. Comparison of GAA values with falling number values have been made by D'Appolonia et al. (1980), Kruger and Tipples (1981), and O'Connell, Rubenthaler, and Murbach (1980), and with amylograph peak viscosity values by D'Appolonia et al. (1980), and Kruger and Tipples (1981). An inter-laboratory evaluation of the GAA by Osborne et al. (1981) showed excellent reproducibility of α-amylase values among three instruments in three different laboratories for the samples tested.

Use of the GAA for commercial α-amylase concentrates (Ranum and Campbell 1981), showed differences in α-amylase activity and stability in dilute extracts prepared from one bacterial concentrate but no differences were found in extracts from a second bacterial concentrate or from the fungal or cereal concentrates studied.

The purposes of this study were to: (1) extend use of the GAA to yeast dough and bread; (2) develop the necessary enzyme extraction procedures; and (3) determine how these procedures would affect the measurement of α-amylase activity.

MATERIALS AND METHODS

Nephelometric Method. The standard AACC Nephelometric Method (AACC method 22-07) was used for all samples.

Falling Number System. The standard AACC Falling Number Method (AACC method 56-81B) was used for unsupplemented flours and those supplemented with cereal or bacterial concentrates.

Wheat Flours. Flours used were a commercially-milled bread flour, unsupplemented, Falling Number (FN) 379, and supplemented with malt at the mill, FN 297; and a commercially-milled all-purpose flour, very low in α-amylase activity, FN 518.

Enzyme Concentrates. The α-amylase concentrates used were those commonly added to flour at the mill or bakery (Table 1).

Preparation of Sample Extracts. The extraction method for flour (AACC method 22-07) was not vigorous enough to disperse yeast dough into a homogeneous slurry, therefore the procedure was modified as summarized in Table 2. Because yeast dough is cohesive, the blade action of a blender was required to disintegrate it. Blending required an increased volume of extracting solution, and thus sample size for effective blade action. Dilution of the viscous slurry from baking dough was necessary to facilitate shaking, and centrifuging and filtering produced clear extracts.

Two experiments investigated effects of the modified procedures on the nephelometric measurement of α-amylase activity. In Experiment 1, a factorial experiment was designed to study effects of type of flour, sample size and volume of extracting solution, extraction time, enzyme stability, method of preparing the β-limit dextrin substrate, and mode of the GAA (Table 3). Thirty-two bread flour samples were weighed; sixteen 10 or 20 g samples of unsupplemented flour and sixteen 10 or 20 g samples of supplemented flour. Flour samples were blended with 50 or 100 ml of extracting solution,

Table 1. Type and Amount of Alpha-Amylase Concentrate Used to Supplement Wheat Flour

Enzyme Concentrate	Alpha-Amylase Activity SKB (Approximate per g)	Amount to Supplement Flour at Approx. .26 SKB/g per 400 g g	Falling[a] Number
Malted barley flour[b]	7.76	0.4800	275
Fungal (Aspergillus oryzae)			
Doh-Tone®[c]	5000	0.0150	-
Enzco®[d]	50,000	0.0015	-
Bacterial (Bacillus subtilis)			
Miles HT-1000™[e]	4700	0.0229	229
Fresh-N®[f]	202	1.2500	247

[a]Average of readings from two samples.
[b]Cargill Company
[c]Pennwalt Corp.   [e]Miles Laboratories
[d]Enzyme Development Corp.   [f]GB Fermentation Industries, Inc.

Table 2. Preparation of Alpha-Amylase Extracts for Analysis with the Grain Amylase Analyzer

Item	AACC Method 22-07	Experiment 1		Experiment 2	
Sample weight	4 g	10 g	20 g	10 g	20 g
Volume extracting solution	20 ml	50 ml	100 ml	50 ml	100 ml
Agitation	Shake 5 min	Blend 5 min		Blend 2 min Dilute slurry Shake 3 min Centrifuge	
Filter	No. 2 Whatman filter paper	No. 2 Whatman filter paper		Filter supernatant No. 5 Whatman filter paper	

Table 3. Effects of Factors Related to Extract Preparation and Measurement of Alpha-Amylase Activity with the GAA, Experiment 1

Factor	F-Value
Wheat flour, unsupplemented or supplemented	2799.00**
Sample weight, 10 g or 20 g	0.18
Extraction time, 30 min or 45 min	0.51
Stability of enzyme, after extraction or refrigeration	1.44
Substrate preparation, day used or day before used	8.41**
Mode of GAA, 1-amylase or 2-amylase	343.60**

**Highly significant ($F_{.01}$ (1,39) = 7.33)

0.5% NaCl and 0.02% $CaCl_2 \cdot 2H_2O$, in a blending jar (approximately 200 ml capacity) on an Osterizer Galaxie Cycle Blend blender at the "Grate" speed. Blending 5 sec was alternated with 5 sec off for the first minute, then for each of the next four min, 30 sec off was alternated with 30 sec of blending at the rate of 5 sec on, 5 sec off. The blended slurry for each sample was filtered for 30 or 45 min.

Experiment 2 was designed because Ranum and Campbell (1981) had reported that α-amylase activity was unstable in extracts from one bacterial enzyme concentrate. In Experiment 2 the stability of α-amylase activity was studied in a factorial experiment using unsupplemented all-purpose flour and the same flour supplemented in the laboratory at recommended levels with malt, fungal or bacterial enzyme concentrates (Table 1). Twelve samples of unsupplemented or supplemented flour were weighed; three were 10 g and three were 20 g samples for each flour. The flour samples were blended with 50 or 100 ml of extracting solution on the "Grind" setting. Blending 5 sec was alternated with 5 sec off for the first 2 minutes of extraction. The slurry was quantitatively transferred from the blender jar to a 300 ml Erlenmeyer flask with 50 or 100 ml additional extracting solution. This step was added to thin the viscous slurries from the baking dough so they could be shaken. The stoppered flask was shaken on an automatic shaker at high speed for 3 minutes. Centrifuge tubes of approximately 50 ml of the slurry from each sample were centrifuged by bringing them up to 10,000 rpm in 4-5 minutes at 50 percent of power, then allowing the centrifuge to stop after being turned off (approximately 9-10 min). The supernatant was poured into a filter paper-lined funnel, the first few drops of filtrate were discarded, and then the sample was collected. Aliquots of the extracts were analyzed with the GAA at 1 h and 1-1/2 h from the start of blending and after 24 h of refrigeration.

Substrate Preparation. Since measurements of α-amylase activity using substrate prepared on the same day were extremely variable in preliminary experiments, the factor of substrate preparation was included in Experiment 1. The β-limit dextrin substrate was either prepared and used the same day, or it was prepared one day, held over night in a beaker at room temperature, and used the next day. The supernatant was carefully pipetted from the layer that formed in the bottom of the beaker and was thoroughly mixed before being used. In Experiment 2, the only substrate used was that prepared the day prior to use.

Analysis of Extracts. In Experiment 1, duplicate 200 µl aliquots of extracts were read on the GAA in the 1-amylase mode, and duplicate 50 µl aliquots were read in the 2-amylase mode. In Experiment 2, duplicate 200 µl aliquots were read in both modes.

Statistical Analysis. Factorial designs used in the two experiments were analyzed using IVAN, an interactive statistical computer program developed at the University of Minnesota (Weisberg and Koehler 1979). In Experiment 1, seven factors, each at two levels, were analyzed. These were flour type, sample weight, extraction time, enzyme stability, substrate preparation, mode of the GAA, and readings for each extract (Table 3). In Experiment 2, five factors, four at two levels and one at three levels, were included in the analysis (Table 4). Factors at two levels were: flour type, sample weight, mode of the GAA, and readings for each extract. The factor at three levels was enzyme stability.

RESULTS AND DISCUSSION

Findings in both experiments seemed to indicate that factors related to modifications in the extraction procedure had little effect on α-amylase activity in extracts from flour. However, factors related to measurement of the enzyme activity appeared to have significant effects on the GAA readings.

Experiment 1. Based on the analysis of variance, a highly significant difference was found between the enzyme activity in the extracts from unsupplemented and supplemented flours as would be expected when enzyme concentrate is added to flour (Table 3).

Factors that did not significantly affect α-amylase activity in the extracts were those related to modifications in the extraction procedure (Table 3). This would indicate that when the one-to-five ratio of sample to extracting solution is maintained, samples of either 10 or 20 g of flour can be used. Since the activity appears to be stable, samples can be extracted for either 30 or 45 min and extracts can be read either immediately after extraction or after 24 h of refrigeration.

Factors that did affect measurement of the enzyme activity were related to substrate preparation and mode of the GAA (Table 3). A highly significant difference was found between readings with freshly prepared substrate and that prepared the day prior to use. Readings for both preparation methods were more variable in the 1-amylase mode than in the 2-amylase mode. Readings in the 1-mode with substrate prepared the day prior to use were slightly less variable than those with substrate prepared the same day. In the 2-mode, readings with both substrates had about the same variability. This would suggest that substrate prepared the day prior to use would be best for the 1-amylase mode, and substrate prepared by either method could be used for the 2-mode. Mode of the GAA had a highly significant effect on readings for the extracts. Mean GAA readings for both the unsupplemented and supplemented flours were higher in the 1-mode than the 2-mode.

Experiment 2. As in Experiment 1, the difference between enzyme activity in extracts from unsupplemented and supplemented flour was highly significant (Table 4).

Factors related to the modified extraction procedure did not seem to affect α-amylase activity in the extracts. Since the sample weight effect was not significant, samples of either 10 or 20 g of flour supplemented with any of the concentrates would be appropriate to use. The enzyme activity in extracts read after the 1 or 1-1/2 h extraction or following refrigeration was not significantly different for flour supplemented with three concentrates, and was barely significant for flour supplemented with two concentrates. Changes in enzyme activity and stability observed by Ranum and Campbell (1981) in dilute extracts from enzyme concentrates apparently were not a serious problem in these flour extracts, thus they could be read after any of these time periods.

Mode of the instrument had a highly significant effect on readings for the flour extracts (Table 4). As in Experiment 1, mean readings for the unsupplemented flour and flour supplemented with each of the concentrates were higher in the 1-amylase than in the 2-amylase mode. Measurements in the 2-mode were less variable than in the 1-mode.

The only highly significant interactions in both experiments occurred between flour and mode of the instrument. They indicated that the two modes of the GAA do not give equivalent values and that one or the other should be used consistently. The 2-amylase mode would be recommended for wheat and wheat products because the measurements are the least variable.

Table 4. Effects of Factors Related to Extract Preparation and Measurement of Alpha-Amylase Activity With The GAA, Experiment 2

		F-Values				
			Type of Concentrate			
		Malted	Fungal		Bacterial	
Factors	D/F	Barley Flour	Doh-tone®	Enzco®	Miles™ HT-1000	Fresh-N®
Flour	1	865.10**	2141.00**	1367.00**	3426.00**	186.00**
Sample weight, 10 g or 20 g	1	1.38	2.73	4.30	3.62	1.39
Stability of enzyme, after 1 h, 1-1/2 h, or refrigeration	2	5.64*	3.24	0.84	5.01*	2.70
Mode of GAA	1	31.29**	94.65**	28.10**	80.04**	11.93**

*Significant ($F_{.05}$ $(2,9) = 4.26$)
**Highly Significant ($F_{.01}$ $(1,9) = 10.56$)

CONCLUSIONS

The α-amylase activity in extracts from flour did not appear to be affected by factors related to modifications in extraction procedures; however, method of substrate preparation and mode of the GAA did affect the measurement of the enzyme activity. These findings were used to develop a procedure for studying changes in α-amylase activity in fermenting and baking yeast dough and bread. In this new procedure, a 20 g sample of flour is blended for two minutes with 100 ml of extracting solution. The slurry is diluted with an equal volume of extracting solution and shaken for three minutes, followed by centrifuging and filtering as described in Experiment 2. After a 1-1/2 h extraction time, the extracts are read in the 2-amylase mode of the GAA using substrate prepared the day prior to use.

NOTE

Published as Paper No. 12271 of the Scientific Journal Series of the Minnesota Agricultural Experiment Station on research conducted under Minnesota Agricultural Experiment Station Project No. 18-015.

Mention of trademark or brand name does not constitute an endorsement and does not imply it is the only product suitable for this application.

ACKNOWLEDGEMENTS

We wish to thank Pennwalt Corp., Ross Milling Division, Cargill Company, Enzyme Development Corp., Miles Laboratories, and GB Fermentation Industries, Inc. for enzyme samples; Pillsbury, Inc.

and Bay State Milling for flour samples; Lucille Caldecott and
Nancy Johnson for technical assistance; Dr. Sanford Weisberg,
Associate Professor, Department of Applied Statistics, University of
Minnesota, for statistical assistance; and Pennwalt Corporation for
loan of the Perkin-Elmer Model 191 Grain Amylase Analyzer.

LITERATURE CITED

American Association of Cereal Chemists. Approved methods of the
    AACC. Method 56-81B approved November 1972; Method 22-07
    approved October 1981.

Anon. 1979. Model 191 Grain Amylase Analyzer. Operating
    Directions C099-1000. Rev. A. Coleman Instruments Division,
    Oak Brook, IL.

Campbell, J. A. 1980a. A new method for detection of sprout
    damaged wheat using a nephelometric determination of alpha-
    amylase activity. Cereal Res. Comm. 8:107-113.

Campbell, J. A. 1980b. Measurement of alpha-amylase in grains.
    Cereal Foods World. 24:46-49.

D'Appolonia, B. L., MacArthur, L. A., Pisesookbunterng, W., and
    Ciacco, C. F. 1980. The use of falling number, amylograph
    and Grain Amylase Analyzer to measure amylase activity. Paper
    read at 65th Annual Meeting, American Association of Cereal
    Chemists, 21-25 September 1980, Chicago, IL.

Kruger, J. E., Ranum, P. M., and MacGregor, A. W. 1979. Note on
    the determination of alpha-amylase with the Perkin-Elmer Model
    191 Grain Amylase Analyzer. Cereal Chem. 56:209-212.

Kruger, J. E. and Tipples, K. H. 1981. Modified procedure for use
    of the Perkin-Elmer Model 191 Grain Amylase Analyzer in deter-
    mining low levels of alpha-amylase in wheats and flours.
    Cereal Chem. 58:271-274.

O'Connell, B. T., Rubenthaler, G. L., and Murbach, N. L. 1980.
    Evaluation of a nephelometric method for determining cereal
    alpha-amylase. Cereal Chem. 57:411-415.

Osborne, B. G., Douglas, S., Fearn, T., Moorhouse, C., and
    Heckley, M. J. 1981. Collaborative evaluation of a rapid
    nephelometric method for the measurement of alpha-amylase in
    flour. Cereal Chem. 58:474-476.

Prasad, K., Watson, C. A., and Carney, Jr., J. B. 1979. Rapid
    nephelometric determination of alpha-amylase activity in
    sprouted wheat kernels. Cereal Chem. 56:43-44.
    Addendum: Prasad, K., Watson, C. A., and Campbell, J. A.
    Cereal Chem. 56:122.

Ranum, P. and Campbell, J. 1981. Nephelometric determination of
    alpha-amylase in enzyme concentrates. Paper read at 66th
    Annual Meeting, American Association of Cereal Chemists, 25-29
    October 1981, at Denver, CO.

Weisberg, S. and Koehler, K. J. 1979. IVAN User's Manual.
    Version 2.1. University of Minnesota, School of Statistics,
    Technical Report No. 266.

Zinterhofer, L., Wardlaw, S., Jatlow, P., and Seligson, D. 1973.
    Nephelometric determination of pancreatic enzymes. 1. Amylase.
    Clinica Acta. 43:5-12.

Environment, Developing Barley Grain, and Subsequent Production of Alpha-Amylase by the Aleurone Layer

P. B. Nicholls, *Waite Agricultural Research Institute, The University of Adelaide, Glen Osmond, Australia*

SUMMARY

Two physiological types of barley grains have been found which differ in their response to the required stimulus for the production of α-amylase by the aleurone layer. Type A grains, the normal, produce α-amylase after treatment with gibberellic acid, and Type B, the atypical, only require water. The latter is found when barley ears are grown under high-intensity sodium lamps. When grown under other light sources, Type B grains are found in 'Clipper' ears grown at >21°C and harvested at the end of grain filling, and in 'Himalaya' ears grown at >24°C and harvested from the end of grain filling. Otherwise the typical variant (Type A) is found. The non-requirement for applied gibberellic acid in Type B endosperm halves did not appear to be due to high residual amounts of endogenous gibberellin-like substances.

This environmentally induced transformation of the aleurone layer may have significant consequences in the understanding and control of preharvest spoilage.

INTRODUCTION

It has been suggested that the concept of the regulation of the mobilization of cereal endosperm reserves by the germinating embryo acting through the hormonal control of the aleurone layer is supported by convincing evidence (Jacobsen *et al.*, 1979). However, some harvests (1969 and 1976) of Himalaya grain in the U.S.A. yielded endosperm halves which produce α-amylase in the absence of exogenously applied gibberellic acid (GA$_3$) (Firn and Kende 1974, Schroeder and Burger 1978), and similarly, Conquest barley endosperm halves have been shown to produce α-amylase I in the absence of GA_3 (MacGregor 1976).

Harvest to harvest variation in the sensitivity of Zephyr and Kenia barley endosperm halves to applied GA_3 has been shown (Jackson 1971). The ability of wheat endosperm halves to produce α-amylase is influenced by (a) the rate of loss of water (Nicholls 1979, King and Gale 1980) and (b) temperature during grain drying (Nicholls 1980). These reports indicate that the production of α-amylase by endosperm halves, either in the presence of GA_3 or in its absence, may be dependent on the environment in which the grains are grown.

This paper is a report on the influence of the effects of temperature and light quality during grain growth on the subsequent potential of endosperm halves to produce α-amylase either in the presence of GA_3 or its absence.

DEFINITIONS AND METHODS

Type A barley grains have an obligate requirement for GA_3 to be present in the incubation medium to initiate production of α-amylase by the aleurone layer.

Type B barley grains will produce this enzyme following the addition of control incubation medium to endosperm halves. In some samples a response to high doses of exogenous GA_3 is observed.

No other biochemical differences have been observed between the two types.

Experiment 1. Three days after anther protrusion Clipper and Himalaya barley plants were transferred from a controlled temperature glasshouse to controlled environment cabinets set at constant temperatures and lit by fluorescent and incandescent lamps 15h/day (see Fig. 1 for spectral distribution) at an irradiance flux of 300 $\mu E.m^{-2}.s^{-1}$. PAR at ear height. Ears were sampled at regular intervals and dried in paper envelopes at 20°C in a dark room. Two grains from each ear were taken for fresh and dry weight measurements and the remainder used in half grain incubates.

Experiment 2. Several days before ear emergence Clipper plants were transferred from a glasshouse to a controlled environment room with alternating day/night temperatures and lit with high pressure sodium lamps and either fluorescent or incandescent lamps for 14h/day (see Fig. 1 for spectral distribution) at an irradiance flux of 750 to 1150 $\mu E.m^{-2}.s^{-1}$. PAR at ear height. The temperatures at night were 5°C lower than during the day. Ears were sampled daily from awn emergence +1 day to +20 days for fresh and dry weight measurements (6 grains/ear) and then at +28 days for GA_3 sensitivity assays. These latter ears were dried in the same manner as those in experiment 1.

Anthesis occurred between 0 and 4 days, and anther protrusion occurred between 4 and 10 days after awn emergence
as the temperature was decreased from 30/25°C to 15/10°C
(Figure 1).

Experiment 3. After transference (as above) Clipper
plants were given either a short day treatment (SD) of
high pressure sodium and fluorescent lamps at 24°C for
10h per day and a night temperature of 19°C, or a long day
treatment (LD) which was achieved by supplementing each
end of the day for 2h with incandescent lamps.

All incubations and α-amylase assays as described by
Nicholls (1980).

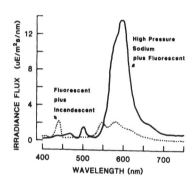

FIGURE 1. Spectral distribution curves for the light sources used in
experiments 1 and 2.

RESULTS AND DISCUSSION

Influence of Environmental Variables on Grain Growth

Experiment 1. In Clipper, the mean rate of dry
weight increase was 2.4 mg.d^{-1}, averaged over all temperatures, and the end of grain filling was adjudged to occur
at 16, 21, 23, 25, 27, and 29 days after anther protrusion
for the respective temperature treatments 27, 24, 21, 18,
15, and 12°C. In Himalaya, the corresponding data are
2.0 mg.d^{-1} and the ends of grain filling occurred 22 to
23 days for all temperature treatments.

Experiment 2. Two phases of increasing dry weight
were distinguished. The first (logarithmic phase) occurred from awn emergence to anther protrusion, i.e., when
the elongating ovary pushed the spent anthers out between
the lemma and palea. It was observed that ovary elongation ceased at this time. The relative growth rates
increased progressively from 0.30 at 15/10°C to 0.58 at
24/19°C and 0.60 g.g^{-1}.d^{-1} at 27/22°C. The second
(linear) phase of growth extended from anther protrusion
to the cessation of grain filling and the growth rates

increased from 1.8 at 15/10°C to a mean value of 2.8 mg.d $^{-1}$ for the 21/16, 24/19, and 27/22 °C treatments.

Experiment 3.   The relative growth rates of the ovary prior to anther protrusion were 0.46 (SD) and 0.45 (LD) g.g^{-1}.d^{-1} and after anther protrusion the growth rates were 2.0 (SD) and 2.4 (LD) mg.d^{-1}.   These values were not significantly less than those in experiment 2 at 24/19°C.

Influence of Environmental Variables on Grain Type

Experiment 1.   Type B grains were only found in ears of Clipper harvested within 6 days from the observed cessation of grain filling and then only when growth temperatures were greater than 21°C (Fig. 2A).   For Himalaya, all ears harvested from the observed cessation of grain filling to harvest ripeness and grown at 27°C were Type B (Fig. 2B).   All ears grown at 12, 15 and 18°C were Type A;   this data is not shown for reasons of clarity.   The apparent reversal of grain type found in Clipper, as grain ripening occurred, may be the basis of the lack of Type B grains in field samples of this culti-var, while Type B grains have been found in field samples of Himalaya (1969 and 1976 harvests).

Experiment 2.   In contrast to the above findings, Type B grains were found in Clipper at all temperatures tried from 15/10 to 27/22°C.   α-Amylase activities in the absence of GA$_3$ varied from 10.4 at 24/19 to 23.3 enzyme units (Eu) at 27/22°C.   In the presence of 10^{-6}M GA$_3$ they varied from 15.0 at 21/16 to 20.9 Eu at 27/22°C.   Neither the -GA$_3$ nor the +GA$_3$ data were correlated with growth temperature.

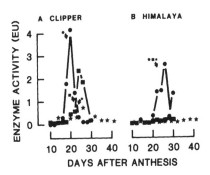

FIGURE 2. α-Amylase activities of -GA$_3$ incubates of endosperm halves taken from (A) Clipper, (B) Himalaya ears, grown at either 27 (●), 24 (■), or 21°C (★).   Grain age at end of grain filling is indicated by an arrow and appropriate symbol.

Experiment 3. The effect of ovary age at harvest was investigated and the data for the -GA₃ incubates are shown in Figure 3. Those data for +GA₃ incubates are not shown as they were found to be similar to those in experiment 2. There appeared to be no significant effect of day length on the light-induced production of α-amylase, but allowing the ears to fully ripen appeared to cause a reduction in this ability.

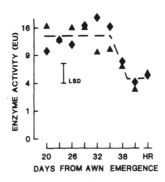

FIGURE 3. α-Amylase activities of -GA₃ incubates of endosperm halves of Clipper ears harvested at various times after ear emergence until harvest-ripe (HR). Symbols: LD (▲), SD (◆). Vertical bar indicates LSD (P = 0.05) between harvest means.

It may be argued that the environments used in these experiments resulted in the presence of high levels of endogenous gibberellin-like substances in imbibed Type B endosperm halves. Attempts to demonstrate the presence of these substances by diffusion (Nicholls 1982) or by extraction experiments (Table 1) were unsuccessful.

TABLE 1
Estimates of amounts of gibberellin-like substances (p.moles GA₃ equivalents/endosperm half) in acid ethyl acetate fractions of homogenates of incubates of either Type A or Type B endosperm halves, together with the observed α-amylase activity of each homogenate

Incubation Time	Enzyme Activity		GA-like Substances	
	Type A	Type B	Type A	Type B
0 h	<0.01	<0.01	0.3	0.4
24 h	0.075	10.5	0.7	0.2

The amounts of these substances found in Type B endosperm halves were neither greater than in Type A halves nor in sufficient quantity to account for the observed production of α-amylase in the absence of the applied hormone.

Alternatively, the broad band yellow irradiation (Fig. 1), which is greater than that found in full sunlight, may have induced a biochemical change in the aleurone equivalent to that induced by 10^{-6} M GA_3. This change is expressed only when a second process, which is induced by detachment of the ears or grains from the plant (Evans *et al.* 1975, Nicholls 1979, 1980, King and Gale 1980) has been at least partially completed; similarly for grains grown at high temperatures and harvested at the appropriate stage of development (Fig. 2). For both grain types, it is hypothesised that this second process sensitizes the aleurone to respond to the initial stimulus irrespective of whether it is GA_3 or the environment.

These data, together with the observed additivity of the effects of light quality and/or exogenous GA_3-induced production of enzyme in some samples of Type B endosperm halves (Nicholls 1982), strongly suggest that the growth environment and GA_3 independently regulate a subsequent key step in the initiation of *de novo* enzyme synthesis, e.g., a cytoplasmically located second messenger.

CONCLUSIONS

It is suggested that the concept of the mobilization of endosperm reserves of barley, controlled by the germinating embryo acting through a hormonal mechanism, should be modified. The data in this paper show that if the grains are grown in certain environments then the addition of water to endosperm halves is a sufficient stimulus to initiate the production of α-amylase, while in other growth environments GA_3 has to be applied.

These results suggest that preharvest spoilage may occur even when the embryo is dormant and unable to produce the appropriate hormonal stimulus for the release and production of hydrolases by the aleurone, i.e., when water is a sufficient stimulus. Further, the observed differences between Himalaya and Clipper suggest that there may be genotypically distinct responses to the environmentally induced changes in grain type and these may be different from the genetically controlled differences in sensitivity to gibberellins in cereals.

ACKNOWLEDGEMENTS

The author acknowledges with thanks financial support from the Barley Improvement Trust Fund and its successor, the Barley Industry Research Council, and CSIRO Phytotron, Canberra, for providing growth facilities.

REFERENCES

Evans, M., Black, M. and Chapman, J. (1975) Induction of hormone sensitivity by dehydration is one positive role for drying in cereal seed. Nature (Lond.) 258, 144-145.

Firn, R.D. and Kende, H. (1974) Some effects of gibberellic acid on the synthesis and degradation of lipids in isolated barley aleurone layers. Plant Physiol. 54, 911-915.

Jackson, D.I. (1971) Factors affecting response in the barley endosperm bioassay for gibberellins. J. Exp. Bot. 22, 613-619.

Jacobsen, J.V., Higgins, T.J.V. and Zwar, J.A. (1979) Hormonal control of endosperm function during germination. In "The Plant Seed" edited by Rubenstein, I., Phillips, R.L., Green, E.C. and Gengenbach, B.C., Academic Press, N.Y. pp. 241-262.

King, R.W. and Gale, M.G. (1980) Preharvest assessment of potential α-amylase production. Cereal Res. Commun. 8, 157-165.

MacGregor, A.W. (1976) A note on the formation of α-amylase in de-embryonated barley kernels. Cereal Chem. 53, 792-796.

Nicholls, P.B. (1979) Induction of sensitivity to gibberellic acid in developing wheat caryopses: effect of rate of desiccation. Aust. J. Plant Physiol. 6, 229-240.

Nicholls, P.B. (1980) Development of responsiveness to gibberellic acid in the aleurone layer of immature cereal caryopses: effect of temperature. Aust. J. Plant Physiol. 7, 645-653.

Nicholls, P.B. (1982) Influence of temperature during grain growth and ripening of barley on the subsequent response to exogenous gibberellic acid. Aust. J. Plant Physiol. 9, (in press).

Schroeder, R.L. and Burger, W.C. (1978) Development and localization of carboxypeptidase activity in embryo-less barley half kernels. Plant Physiol. 62, 458-462.

Observations on the Development of the Testa and Pericarp in Barley

M. P. Cochrane and C. M. Duffus, *School of Agriculture, University of Edinburgh, Edinburgh, Scotland*

SUMMARY

The distribution of cuticular layers and the accumulation of phenolic substances in developing barley caryopses are described and discussed in relation to the uptake of water and oxygen prior to sprouting.

INTRODUCTION

When Krauss (1933) presented her detailed study of the development of the outer layers of barley and wheat caryopses, she paid particular attention to the cuticular layers and interpreted them as structures which control the uptake of water and solutes. Morrison (1975) examined the ultrastructure of the cuticular layers of wheat caryopses and established that the innermost cuticular layer is derived from the nucellar epidermis. A similar interpretation of the origin of the innermost cuticular layer of the barley caryopsis was subsequently given by Cochrane and Duffus (1979). At the same time it was suggested that cuticular layers restrict the movement of atmospheric oxygen into the caryopsis and of photosynthetically-derived oxygen out of the pericarp. It has long been recognised that white-grained wheats are more susceptible to pre-harvest sprouting than red-grained wheats. Using scanning electron microscopy it has been shown that in the testa of cultivars which are resistant to sprouting there is a higher concentration of sulphur than in the testa of cultivars which are susceptible. It was suggested that the type of protein present in the testa affects the permeability of this tissue to oxygen and hence the resistance of the cultivar to pre-harvest sprouting (Belderok, 1976).

This paper presents some further observations on the structure of the pericarp and testa of developing barley grains, particular attention being paid to features which may influence the supply of oxygen to the embryo.

MATERIALS AND METHODS

Caryopses were harvested from barley cv. Julia grown in a controlled environment chamber at 20°C in a 16 h day, and from cv. Midas grown in a glasshouse (Cochrane and Duffus, 1979). The age of each caryopsis was estimated on a developmental time scale of 60 days' from anthesis to harvest ripeness. Caryopses were fixed in 2.5% glutaraldehyde in 0.025 M phosphate buffer pH 7.2 and then either post-fixed in 2% osmium tetroxide, dehydrated in an acetone series and embedded in epoxy resin, or dehydrated in an ethanol series and embedded in LR White resin. For electron microscopy, sections were stained in uranyl acetate followed by lead citrate. Epoxy resin sections 2 μ thick were stained for light microscopy using toluidine blue at pH 9.5. Caropyses embedded in LR White resin were sectioned at 2 μ and stained using Sudan Black B, ruthenium red, toluidine blue at pH 4.4, or the Hoepfner-Vorsatz method for locating phenolic substances (Ling-Lee *et al.*, 1977). To prepare whole mounts of various tissue layers, caryopses were fixed in 2.5% formaldehyde overnight, leached using ethanol at 60°C and cleared in 70% lactic acid at 60°C.

OBSERVATIONS

Pericarp

The pericarp of a developing barley grain is bounded by an epidermis with walls which are slightly thickened and stain darkly in toluidine blue. Over most of the caryopsis the chlorophyll-containing cross cells are tightly packed and are very uniform in shape, being elongated in a plane at right angles to the long axis of the grain (Plate 1a). However, over the developing embryo the chlorophyll-containing cells are irregular in shape and the tissue resembles the spongy mesophyll of a leaf (Plate 1b). The cuticular layer on the outer wall of the outer epidermis is much thicker at the middle of the grain than on the part of the pericarp which covers the embryo (Plate 1c, d). The pericarp is only a few cells thick in the micropylar region (Plate 1e) and these cells become crushed as the embryo grows. Close to the junction of the pericarp and the rachilla the epidermis is not uniform. Some cells have wall thickening which does not stain in toluidine blue at pH 9.5. Only a small zone of loosely-packed tissue separates these cells from the ovule. On the rachilla side of the unstained cells there are cells which stain darkly in toluidine blue at pH 9.5 and next to these is the lodicule which stains pink in ruthenium red and toluidine blue at pH 4.4.

Testa

The testa is composed of two layers of cells which differ considerably from each other throughout development. In the 22-'day' caryopsis, cv. Julia, shown in Plate 2a the cells of the outer layer have dense cytoplasm and few vacuoles. In contrast, the cells of the inner layer have many vacuoles of all sizes, and cytoplasm which is aggregated, leaving large electron-lucent areas. The radial walls are characteristically buckled. In transverse sections of older caryopses the vacuoles contain large, lenticular, electron-dense deposits (Plate 2b). This material is more osmiophilic than lipid bodies and is similar to the deposits observed in the vacuoles of the cells of the chalazal region. In sections of LR white-embedded caryopses stained with toluidine blue at pH 4.4, cell contents both in the testa and in the chalazal region are green, and when stained using the Hoepfner-Vorsatz reaction they are yellow/orange turning to rusty red or tan colour as the age of the caryopsis increases.

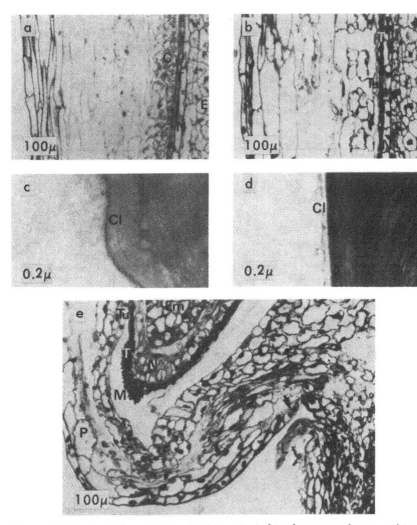

Plate 1(a) (b) L.S. of dorsal side of 23-'day' caryopsis cv. Midas embedded in epoxy resin stained in toluidine blue pH 9.5.
(a) mid-grain   (b) base of grain

(c) (d) T.S. of outer wall of pericarp epidermis of 50-'day' caryopsis cv. Midas, embedded in LR White resin (see text).   (c) mid-grain
(d) over embryo.

(e) L.S. base of 23-'day' caryopsis cv. Midas, embedded in epoxy resin.

C, cross cells;   E, endosperm;   Em, embryo;   L, lodicule;
M, micropylar region;   N, nucellus;   P, pericarp;   T, testa;
Tu, tube cells;   arrow, pericarp epidermal cells with unstained walls.

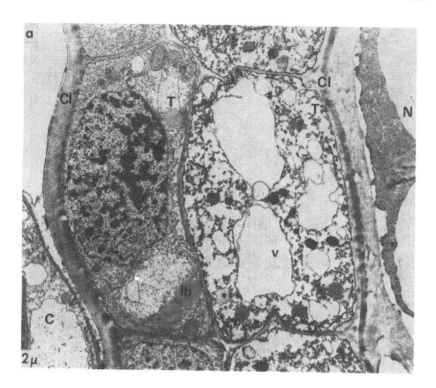

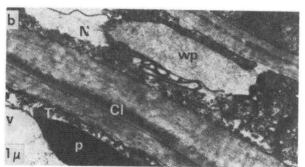

Plate 2 (a) T.S. mid-caryopsis, cv. Julia, 20-'day', grown at 20°C, embedded in epoxy resin.

(b) T.S. mid-caryopsis, cv. Midas, 28-'day', embedded in epoxy resin.

Cl, cuticular layer; lb, lipid body; p, phenolic material; v, vacuole; wp, wall proliferation. (See also Plate 1.)

The cuticular layer of the testa is continuous over the whole
ovule except in two places - the micropylar region and the junction
with the pericarp at the crease. In the micropylar region the testa
is several cells thick (Plate 1e and Plate 3a, b, c). In sections
stained in toluidine blue at pH 4.4 the contents of the cells occupying the gap at the micropyle stain bright green, indicating the
presence of phenolic substances. This confirmed the observations
made on whole mounts of cleared tissue. The micropylar region of an
18-'day' caryopsis is almost colourless but in older caryopses the
testa cells in the micropylar region have dense brown contents
(Plate 3a). There is little, if any, brown material in the more or
less isodiametric testa cells covering the developing embryo (Plate
3d) but those covering the endosperm contain large amounts of dark
reddish-brown material (Plate 3e).

The other break in the cuticular layer surrounding the testa is
at the junction with the pericarp. This occurs along the length of
the caryopsis on the ventral side. As at the micropyle, the cells
occupying the gap contain phenolic substances. These can be detected
as early as 15 'days' after anthesis and gradually increase in concentration. In cleared tissue the colour changes from light golden
brown to dark reddish-brown between 18 and 48 'days'. It is interesting to note that the contents of the testa cells stain more intensely
at the crease than elsewhere in the caryopsis (Plate 3f). By about
40 'days' the walls of the chalazal cells are lined with material
which has a staining reaction in Sudan Black B similar to that of
suberin in potato tuber epidermis.

DISCUSSION

The observations presented here have confirmed many of those
made by Krauss (1933), Pugh et al., (1932) and Bradbury et al.,
(1956). In addition, the accumulation of phenolic substances in
various tissues has been described.

In a grain detached from the ear, water can enter at the attach-
ment point and rise through the loosely-packed parenchymatous cells
of the pericarp to the two regions where there are gaps in the cuticular layer of the testa. The discontinuity of the thick outer cuticular layer at the micropylar region has been clearly demonstrated.
Pugh et al., (1932) suggested that the inner layer of testa cells
folds back over the outer layer at the micropyle and that the inner
cuticular layer is then continuous with the outer cuticular layer.
Recent work has shown that the inner cuticular layer is part of the
nucellus and not of the testa (Morrison, 1976; Cochrane & Duffus,
1979). Observations on the ultrastructure of this region are needed
to establish the origin and extent of the thin cuticular layer which
was observed on some of the cells of the testa at the micropyle of
barley in this investigation, and of wheat by Bradbury et al., (1956).
During the last third of the period of grain development the cuticular layers of the testa and nucellus appear to be continuous with a
lining of 'suberin' on the walls of the chalazal cells. Nothing is
known of the effect of this layer of 'suberin' on the movement of
water into the endosperm.

In pre-harvest sprouting, grains are attached to the ear, and
water would enter the caryopsis either through the vascular bundles

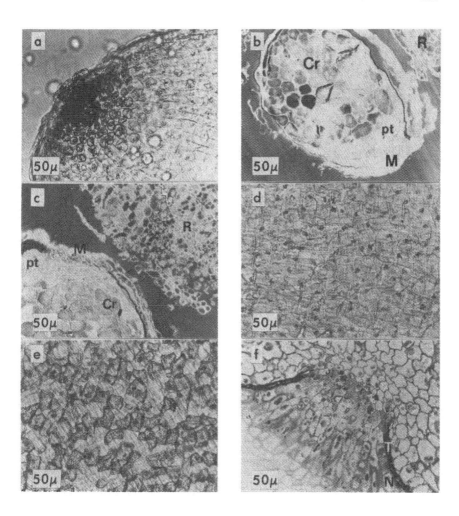

Plate 3 (a) (d) (e) Whole mounts of testa of caryopses of cv. Midas cleared in lactic acid (see text).

(b) (c) (f) T.S. of caryopses of cv. Midas, fixed in glutaraldehyde and embedded in LR White resin.

(a) micropylar region, 33-'day' (b) micropylar region, 42-'day', stained in Sudan Black B (c) micropylar region, 50-'day', stained in Sudan Black B (d) testa over embryo (e) testa on flank at base of grain (f) crease region stained in toluidine blue at pH 4.4.

Cr, coleorhiza; pt, pectinaceous material; R, rachilla, (see also Plates 1 and 2).

of the rachilla or through the epidermis of the pericarp. The pericarp cuticular layer is very thin over the embryo and may be absent
over a small group of cells next the junction with the rachilla.
Lodicules are feathery in form and apparently composed of pectinaceous material. They are joined to the rachilla at the level at
which the micropylar region of the 50-'day' embryo lies close to the
rachilla, and it is likely that they would hold any moisture falling
on that part of the grain and thus increase the amount of water
available for imbibition,

However, water uptake does not necessarily lead to germination
(Takahashi, 1979) and the significance of the testa is thought to
lie in its control of the amount of oxygen available to the embryo
(Roberts, 1962). Since the pericarp epidermal cuticle is very thin
over the embryo, and air spaces are extensive, atmospheric oxygen
would be readily available to the ovule. No direct measurements of
the permeability to oxygen of the cuticular layer of the testa of
barley have been made but it has been assumed to be low on account
of the lipid content of the cuticular material. The cuticular layer
of the testa is well developed very soon after anthesis and so it
would appear that most of the oxygen reaching the developing embryo
must pass through the micropylar region. As the caryopsis develops,
phenolic compounds, probably tannins, accumulate in the vacuoles of
all cells in the inner layer of the testa, except those covering the
embryo. The testa cells at the micropyle and at the crease are
particularly rich in phenolic substances, as are the cells of the
chalazal region. Polyphenols can react with oxygen and can also form
complexes with proteins. As yet there is no direct evidence that the
polyphenolic material in the micropylar region is responsible for
limiting the oxygen available to the mature embryo but it does seem
that the testa must be punctured or the plug of tannin-rich material
at the micropyle must be broken, either physically or by the slower
processes of chemical change, before sufficient oxygen reaches the
embryo for germination to take place. Further information is needed
on the morphology and tannin content of the micropylar region in
sprouting-resistant and sprouting-susceptible cultivars.

ACKNOWLEDGEMENT

This work was supported by a grant from the Agricultural
Research Council, London.

REFERENCES

Belderok, B. (1976). Changes in the seed coat of wheat kernels
    during dormancy and after-ripening. Cereal Res. Comm. 4,
    165-171.
Bradbury, D., MacMasters, M.M. and Cull, I.M. (1956). Structure of
    the mature wheat kernel II. Microscopic structure of pericarp,
    seed coat, and other coverings of the endosperm and germ of
    hard red winter wheat. Cereal Chem. 33, 342-360.
Cochrane, M.P. and Duffus, C.M. (1979). Morphology and ultrastruc-
    ture of immature cereal grains in relation to transport. Ann.
    Bot. 44, 67-72.

Krauss, L. (1933). Entwicklungsgeschichte der Früchte von *Hordeum*, *Triticum*, *Bromus*, und *Poa*, mit besonderer Berücksichtigung ihrer Samenschalen. Jahrb. wiss. Botan. 77, 733-808.

Ling-Lee, M., Chilvers, G.A. and Ashford, A.E. (1977). A histochemical study of phenolic materials in mycorrhizal and uninfected roots of *Eucalyptus fastigiata* Deane & Maiden. New Phytol. 78, 313-328.

Morrison, I.N. (1975). Ultrastructure of the cuticular membranes of the developing wheat grain. Can. J. Bot. 53, 2077-2087.

Pugh, G.W., Johann, H. and Dickson, J.G. (1932). Relation of the semipermeable membranes of the wheat kernel to infection by *Gibberella saubinetii*. J. Agr. Research 45, 609-626.

Roberts, E.H. (1962). Dormancy in rice seed III. The influence of temperature, moisture and gaseous environment. J. Exp. Bot. 13, 75-94.

Takahashi, N. (1980). Effects of the environmental factors during seed formation on pre-harvest sprouting. Cereal Res. Comm. 8, 175-183.

Cereal Endosperm Degradation During Initial Stages of Germination

A. W. MacGregor, *Grain Research Laboratory, Canadian Grain Commission, Winnipeg, Manitoba, Canada*

SUMMARY

Longitudinal sections of barley kernels at different stages of germination were examined by scanning electron microscopy. Degradation of the protein matrix, endosperm cell walls and starch granules started in the ventral crease region of the endosperm-embryo junction and moved along this junction to the dorsal edge. After 72 hr of germination at 18°C there was a narrow band of highly degraded starch granules along the endosperm-embryo junction but only limited degradation was detected under the aleurone. Protein and cell wall degradation had progressed much further into the endosperm.

INTRODUCTION

The plant hormone, gibberellic acid, has been shown to induce the synthesis of a number of hydrolytic enzymes, including α-amylase, in the aleurone cells of cereal grains (Yomo 1958; Paleg 1960; Briggs 1963; MacLeod et al 1964). This finding has led to extensive research on the aleurone layer of barley in particular, and on the role played by this tissue during barley germination and malting. During germination the hormone moves from the embryo into the aleurone layer and, as it moves through this layer towards the distal tip of the grain, it triggers the synthesis of hydrolytic enzymes (Palmer 1980). From such studies has emerged a general picture of endosperm modification during malting in which physical disintegration starts immediately below the aleurone layer close to the embryo and from there moves into the interior of the endosperm and also towards the distal tip of the kernel (Palmer 1980).

However, the embryo is also capable of producing hydrolytic enzymes. Recent findings (Gibbons 1979, 1980; Okamoto et al 1980; MacGregor 1980) have confirmed earlier reports that the embryo is an important source of hydrolytic enzymes during initial stages of germination (Brown and Morris 1890; Dickson and Shands 1941; Briggs 1964). Therefore, there is some disagreement over the relative importance of hydrolytic enzymes produced by aleurone and embryonic

Paper No. 492, Canadian Grain Commission, Grain Research Laboratory, 1404-303 Main Street, Winnipeg, Manitoba, R3C 3G8.

tissues at early stages of germination. The main objective of the present study was to detect and follow the initial modification pattern within barley endosperms using scanning electron microscopy.

MATERIALS AND METHODS

Germination. Kernels of the two-rowed barley Klages were soaked in sodium hypochlorite solution (1.5%) for 20 min and washed with sterile water. Germinations were carried out at 18°C in sterile petri dishes containing three pieces of Whatman No. 1 filter paper, 4 mL of water and 50 kernels. After 24, 48, 72, 96, 120 and 164 hr, samples were removed, cleaned, frozen and freeze-dried.

α-Amylase Activity. Freeze-dried kernels were ground in a Wiley mill through a 1 mm sieve. Ground samples were assayed for α-amylase activity as described previously using amylopectin β-limit dextrin as substrate (Briggs, 1961; MacGregor et al 1971).

Scanning Electron Microscopy. Kernels were cracked open longitudinally through the crease, fixed to microscope stubs with silver paint and coated with gold. Samples were analyzed on a JEOL 35C scanning electron microscope at an accelerating voltage of 10 KV. Photomicrographs were taken on Plus-X pan Kodak film.

RESULTS AND DISCUSSION

The α-amylase activity of Klages barley increased slowly during the first 48 hr of germination but thereafter there was a rapid increase in enzyme activity (Fig. 1). This profile of α-amylase

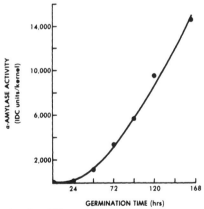

Figure 1. Synthesis of α-amylase in Klages barley during germination.

synthesis illustrates that the enzyme potential of Klages is more typical of six-rowed barley than of two-rowed cultivars (MacGregor 1978). In this report only changes taking place in the endosperm during the first 72 hr of germination will be discussed.

The structure of a portion of the endosperm-embryo junction of a sound kernel of Klages barley is shown in Fig. 2. The scutellar

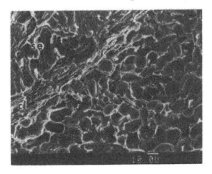

Figure 2. Scanning electron photomicrograph of sound barley kernel - se = scutellar epithelium; cl = crushed layer.

epithelium with its single layer of large columnar cells is readily apparent, as is the layer of crushed cells at the edge of the endosperm. This crushed material is thought to be the remnants of endosperm cells whose contents were depleted during kernel growth and the empty cells were later crushed between the growing embryo and endosperm tissues (Brown and Morris 1890). In the endosperm, starch granules are deeply embedded in a protein matrix. The physical structure of the endosperm appears to vary widely even between kernels of the same barley sample. A thick protein matrix can bury not only starch granules but may also cover cell walls so that individual cells may be difficult to see. In other kernels, where the matrix may be much thinner, much more structure may be apparent.

Because preliminary studies had shown that initial signs of endosperm degradation were detected at the endosperm-embryo junction (MacGregor 1980) particular attention was paid to this area during this study.

After 24 hr of germination all kernels examined contained areas of degraded starch granules (Fig. 3). These were always found in

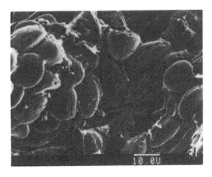

Figure 3. Scanning electron photomicrograph of 24 hr germinated kernel.

areas adjacent to the embryo and usually close to the crease edge of the endosperm. Starch degradation, then, appeared to commence in small pockets at the crease edge of the endosperm adjacent to the embryo. Sometimes small areas of degraded granules interspersed with areas of intact endosperm were observed along the endosperm-embryo junction but rarely were areas of degradation found close to the dorsal edge of the endosperm. Large starch granules were attacked initially at the equational groove and then randomly over the granule surface as described by other workers (Evers and McDermott 1970; Palmer 1972; Maeda et al 1978).

In areas containing degraded starch granules there was extensive breakdown of the protein matrix and cell wall material. This degradation appeared to emanate from the embryo and preceded action of α-amylase. Therefore, β-glucanase and proteolytic activity were detected prior to that of σ-amylase. This agrees with previous studies (Okamoto et al 1980).

After 48 hr of germination large areas of endosperm close to the embryo were virtually free of protein matrix and cell wall material (Fig. 4A). Starch granules close to the embryo were

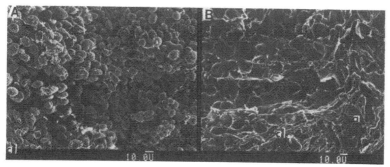

Figure 4. Scanning electron photomicrograph of 48 hr germinated kernel. A - endosperm close to embryo and crease edge; B - endosperm-embryo-aleurone junction at dorsal edge; a - aleurone layer.

extensively degraded. Occasionally some signs of degradation were observed at the dorsal edge of the endosperm at this stage in germination but more often this area was similar to that shown in Fig. 4B. Very little degradation of proteinaceous or cell wall material can be seen and no starch hydrolysis is apparent. At this stage the aleurone layer does not appear to excrete hydrolytic enzymes into the endosperm.

Endosperm degradation was much more extensive after 72 hr. Half-way along the endosperm-embryo junction there was a narrow band of severely destroyed starch granules and above that a broader band of more lightly attacked granules (Fig. 5A). Again, protein and cell wall breakdown had preceded α-amylase attack and had progressed extensively through the proximal end of the endosperm. Although endosperm modification had occurred along the entire length of the endosperm-embryo junction only limited starch degradation was detected at the dorsal edge (Fig. 5B). However, at least some

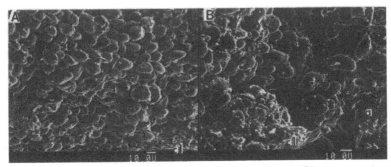

Figure 5.   Scanning electron photomicrograph of 72 hr germinated kernel.   A - half-way along endosperm-embryo junction; B - endosperm-embryo-aleurone junction at dorsal edge.

granule attack was visible at the endosperm-embryo-aleurone corner in most kernels examined. This attack appeared to come from the embryo and not the aleurone layer. There was evidence that in areas further from the embryo, protein and cell wall hydrolysis may have been carried out by enzymes from the aleurone layer.

These results support earlier reports (Brown and Morris 1890; Gibbons 1979, 1980) that during initial stages of germination of barley kernels the embryo and not the aleurone layer is primarily responsible for the synthesis of α-amylase. Similar results have been obtained with rice (Okamoto and Akazawa 1979), wheat, rye, oats and maize (Okamoto et al 1980) showing that this is general for all cereal grains.

Degradation of endosperm protein in germinating barley appears to emanate from the embryo and precedes α-amylase secretion (Okamoto et al 1980). This degradation, presumably, is caused by proteolytic enzymes excreted from the embryo but the possibility of sequential activation of proteases within the endosperm should not be ignored.

Cell wall degradation also precedes starch hydrolysis and this, too, appears to start at the embryo in barley (Gibbons 1980) as well as in wheat (MacGregor and Matsuo 1982). This suggests that in barley, β-glucanase and the recently reported β-glucan solubilase (Bamforth and Martin 1981) are initially synthesized in the embryo. Wheat endosperm cell walls are composed predominantly of pentosans (Mares and Stone 1973) so that pentosanases must originate from the embryo. Again, these results could be explained by sequential acti-vation of cell wall degrading enzymes within the endosperm.

To follow the actual movement of specific enzymes within cereal endosperms during germination, immunochemical techniques such as those used for α-amylase, must be developed. Much more research into embryo metabolism is required to obtain a deeper understanding of the formation and action of hydrolytic enzymes during early stages of germination of cereal grains.

REFERENCES

Briggs, D.E. 1961. A modification of the Sandstedt, Kneen and Blish assay of α-amylase. J. Inst. Brew. 67:427-431.

Briggs, D.E. 1963. Biochemistry of barley germination: Action of gibberellic acid on barley endosperm. J. Inst. Brew. 69:13-19.

Briggs, D.E. 1964. Origin and distribution of α-amylase in malt. J. Inst. Brew. 70:14-24.

Brown, H.T. and Morris, G.H. 1890. Researches on the germination of some of the Gramineae. Part. 1. J. Chem. Soc. 57:458-528.

Dickson, J.G. and Shands, H.L. 1941. Cellular modification of the barley kernel during malting. Am. Soc. Brew. Chem. Proc. 1-9.

Evers, A.D. and McDermott, E.E. 1970. Scanning electron microscopy of wheat starch. II. Structure of granules modified by α-amylolysis – preliminary report. Die Starke 22:23-26.

Gibbons, G.C. 1979. On the localization and transport of α-amylase during germination and early seedling growth of Hordeum vulgare. Carlsberg Res. Commun. 44:353-366.

Gibbons, G.C. 1980. On the sequential determination of α-amylase transport and cell wall breakdown in germinating seeds of Hordeum vulgare. Carlsberg Res. Commun. 45:177-184.

MacGregor, A.W. 1978. Changes in α-amylase enzymes during germination. J. Am. Soc. Brew. Chem. 36:1-5.

MacGregor, A.W. 1980. Action of malt α-amylases on barley starch granules. Tech. Quart. of MBAA 17:215-221.

MacGregor, A.W., LaBerge, D.E. and Meredith, W.O.S. 1971. Changes in barley kernels during growth and maturation. Cereal Chem. 48:255-269.

MacGregor, A.W. and Matsuo, R.R. 1982. Starch degradation in endo-sperms of barley and wheat kernels during initial stages of germination. Cereal Chem. 59:210-216.

MacLeod, A.M., Duffus, J.H. and Johnston, C.S. 1964. Development of hydrolytic enzymes in germinating grain. J. Inst. Brew. 70:521-528.

Maeda, I., Kiribuchi, S. and Nakamura, M. 1978. Digestion of barley starch granules by the combined action of α- and β-amylases purified from barley and barley malt. Agric. Biol. Chem. 42:259-267.

Mares, D.J. and Stone, B.A. 1973. Studies on wheat endosperm. I. Chemical composition and ultrastructure of the cell walls. Aust. J. Biol. Sci. 26:793-812.

Okamoto, K. and Akazawa, T. 1979. Enzymic mechanisms of starch breakdown in germinating rice seeds. 7. Amylase formation in the epithelium. Plant Physiol. 63:336-340.

Okamoto, K., Kitano, H. and Akazawa, T. 1980. Biosynthesis and excretion of hydrolases in germinating cereal seeds. Plant and Cell Physiol. 21:201-204.

Paleg, L.G. 1960. Physiological effects of gibberellic acid. II. On starch hydrolyzing enzymes of barley endosperm. Plant Physiol. 35:902-906.

Palmer, G.H. 1972. Morphology of starch granules in cereal grains and malts. J. Inst. Brew. 78:326-332.

Palmer, G.H. 1980. The morphology and physiology of malting barleys in: Inglett, G.E. and Munck, L., eds. Cereals for Food and Beverages. Academic Press:New York, 301-338.

Yomo, H. 1958. Barley malt. Sterilization of barley seeds and the formation of amylase by separated embryos and endosperms. Hakko Kyokaishi 16:444-448.

The Action of Plant Hormones on Endosperm Breakdown and Embryo Growth During Germination of Barley

Gregory C. Gibbons, *Carlsberg Research Laboratory, Copenhagen Valby, Denmark*

SUMMARY

Germination is the result of a number of physiological responses: endosperm breakdown resulting from the production of hydrolytic enzymes, rootlet growth, and coleoptile growth. The present results report the effects of GA_3, IAA, and ABA on these responses. It is concluded that initial endosperm cell wall breakdown can proceed normally under conditions where rootlet and coleoptile growth are severely inhibited. The implications of these findings to preharvest sprouting are discussed.

INTRODUCTION

Since the introduction of the concept by Gibbons (1980a) at the last Sprouting Symposium that the production of hydrolytic enzymes during germination of barley was not solely the result of a gibberellic acid/aleurone layer response, but rather a substantial initial scutellar mediation of hydrolases followed by a later involvement of the aleurone, it has been necessary to re-examine many aspects of preharvest sprouting. It is apparent from later work (Gibbons 1980b; Gibbons 1981; Munck et al. 1981; Okamoto et al.1980) that in several cereals the scutellum plays an important role in the mediation of the first hydrolases. As in most cases preharvest sprouting is considered to represent the initial physiological steps of germination, it is important to elucidate the control mechanisms involved. Indeed, the demonstration that the scutellum apparently produces alpha-amylase and beta-glucanases under conditions where the aleurone layer is incapable of effecting this task (Chrispeels and Varner 1967; Gibbons 1981) only serves to stress the importance of such further physiological studies of the early stages of germination.

In the present paper the effect of three plant hormones, widely considered of importance in germination, is examined in the light of their effects on both endosperm cell wall breakdown and embryo growth.

MATERIALS AND METHODS

Chemicals

Calcofluor white M2R new (di-sodium 4,4'-bis (4 - anilinobis-2-hydroxy-ethyl amino-S-triazin-2-ylamino)-2,2'-stilbenedisulphonate) was obtained from Cyanamid, U.S.A. via Struers, Denmark. Fast Green F.C.F. (Fast Green 3; lot 17761) was obtained from Hopkin and Williams, Chadwell Heath, Essex, England. 96% ethanol was obtained from De Danske Spritfabrikker, Denmark, and fluorescence-free glycerol was from Merck, West Germany.

Gibberellic acid (GA , grade III, minimum content Gibberellin A3, 90%, lot No. 98c-0008), ± cis-trans abscisic acid (ABA, grade IV, approx. 95%, lot No. 129c-3919) and Indole-3-acetic acid (IAA, lot No. 24c-3320) were obtained from Sigma Chemical Company, St. Louis, MO., U.S.A.

All other reagents were of analytical grade.

Plant Material

Seeds of Hordeum vulgare L. (cv. Nordal) were obtained from the 1981 harvest at the Carlsberg Plant Breeding research station, Allingemaglegard. Seeds were fractionated on a Mini-Petkus laboratory sorting machine (Schule, Hamburg, West Germany) and the fraction retained between 2.85 mm and 3.00 mm was stored at room temperature and used for all experiments. Seeds (30) were pushed into sterile sand (100 g) in a sterile petri dish and 21 mL of the appropriate hormone solution or distilled water were added. The seeds were germinated in a darkened room for the required time at 15°C (98% relative humidity). To increase the permeability of the seeds to exogenous hormones, in the appropriate treatments, seeds were either 1) manually abraded (distally decapitated) by removing the distal 0.5 mm of the seed with a scalpel or, 2) nicked by placing 3 manual incisions (0.5 mm) on the ventral surface of the seed and 2 incisions on the dorsal side.

Endosperm Cell Wall Breakdown Measurements

Endosperm cell wall breakdown was followed with the calcofluor/fast green method described previously from this laboratory (Aastrup et al. 1981). The cell wall breakdown (modification) was assessed visually using a commercial Malt Modification Analyser - system Carlsberg (Carlsberg Research Center, Copenhagen, Denmark).

Experimental Design

The effects of gibberellic acid, indole acetic acid, and abscisic acid alone and in combinations were studied in the following way: whole, distally decapitated or nicked seeds were germinated on sand for 7 days in the presence of hormones or hormone combinations

(Table 1). The experiments were designed in such a way that each treatment/day was contained in individual petri dishes, thus obviating the problem of disturbing root growth during sampling.

TABLE 1. Definition of Treatments

	Control (whole seeds)	Distal Decapitation	Nicked
Control$_5$ (H_2O)	A	H	O
IAA 10^{-5} M	B	I	P
GA$_3$ 10^{-5} M	C	J	Q
IAA 10^{-5} M + GA$_3$ 10^{-5} M	D	K	R
ABA 5 x 10^{-5} M	E	L	S
IAA 10^{-5} M + ABA 5 x 10^{-5} M	F	M	T
IAA 10^{-5} M + GA$_3$ 10^{-5} M + ABA 5 x 10^{-5} M	G	N	U

The results are presented in 6 figures. Figures 1, 2, and 3 (the endosperm cell wall data) demonstrate the span of cell wall modification in individual seeds for each day of germination. The modification class is defined as:

Class 0 : unmodified (no cell wall breakdown)
Class 1 :   0 - 20% modified
Class 2 : 20 - 40% modified
Class 3 : 40 - 60% modified
Class 4 : 60 - 80% modified
Class 5 : 80 - 95% modified
Class 6 : 95 - 100% modified
Class 7 : 100% modified (no cell walls remaining)

Due to difficulties in sectioning overmodified green malt, only data up to day 6 are included in these figures. Figures 4, 5 and 6 are the data for the average rootlet and coleoptile growth during the 7 day time course of germination. Figures for the average endosperm modification on each day are included for each treatment and presented below the columns showing rootlet and coleoptile growth.

RESULTS

Endosperm Cell Wall Breakdown

Distal decapitation and nicking alone had no marked effect on the overall rate of modification of control samples grown in water (Fig. 1). It was found, however, that the modification values for the nicked 4 day germinated seeds were low. Closer examination of all the nicked seed treatments germinated for 4 days (Figs. 1, 2

172

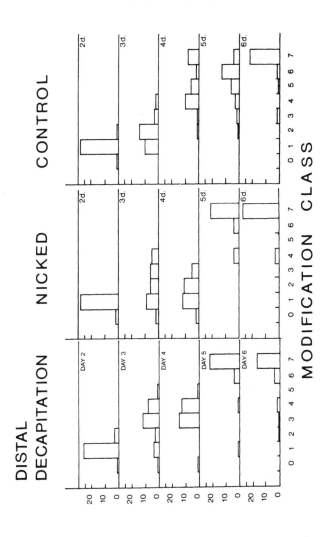

Figure 1.   Endosperm cell wall breakdown. For explanation of treatments see EXPERIMENTAL DESIGN.

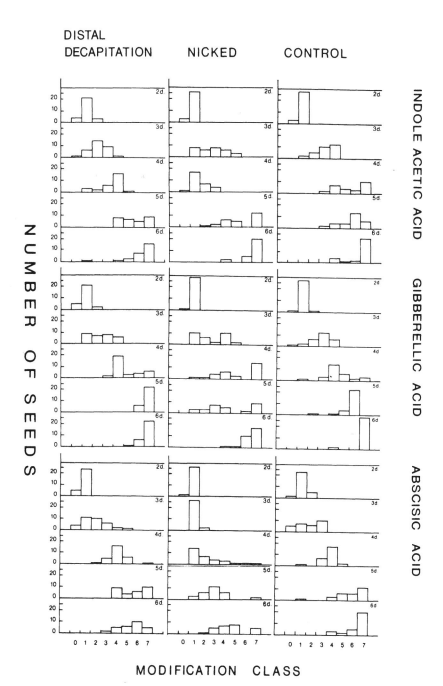

Figure 2. Endosperm cell wall breakdown in the presence of phytohormones. For explanation of
treatments see EXPERIMENTAL DESIGN.

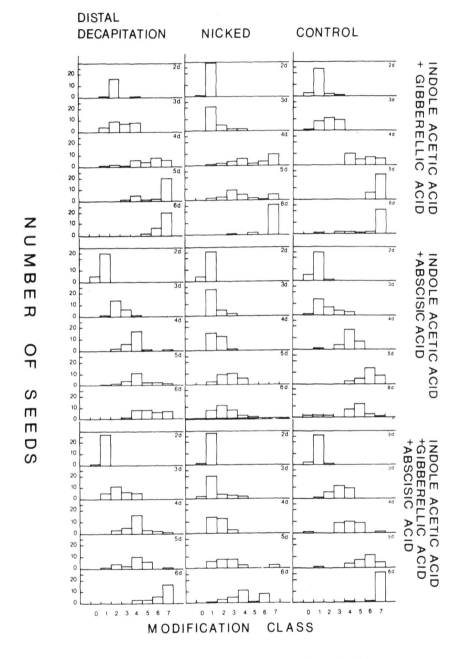

Figure 3. Endosperm cell wall breakdown in the presence of mixture of phytohormones. For explanation of treatments see EXPERIMENTAL
DESIGN.

and 3) indicated that apart from the GA₃ and GA₃ + IAA treatments, all values were lower than expected. As all hormone treatments for 4 day germinated nicked seeds were set to germination and processed as a group, it is suspected that the generally lower modification seen here is the result of an error in temperature control in the early phase of germination.

IAA gave no stimulation of modification in any of the seed treatments (Fig. 2). GA₃ on the other hand, had a marked stimulatory effect between the 3rd and 4th day of germination (Fig. 2). This effect has been documented previously (Gibbons 1980b; Gibbons 1981) and shown to result from a stimulation of hydrolase production by GA₃ in the distally located aleurone cells. The greatest GA₃ effect was seen in seeds where the pericarp and testa had been opened, thereby allowing facilitated access of the hormone to the aleurone layer (i.e., distal decapitation and nicked).

ABA alone had a marked inhibitory effect on the rate of modification of distally decapitated and nicked seeds. This result is in keeping with previous findings where 5×10^{-5} M ABA effectively inhibited GA₃ stimulated production of cell wall breakdown enzymes in barley aleurone layers both in vivo and in vitro (Gibbons 1981; Mozer 1980). In the present experimental series ABA had little effect on the rate of endosperm cell wall breakdown in whole seeds.

When combinations of hormones were used (Fig. 3), it was found that IAA, together with GA₃, gave very similar results to GA₃ alone. A stimulation in the rate of modification between the 3rd and 4th day was observed in all seed treatments. IAA and ABA together showed no early modification effects while in the later stages, the progression of cell wall breakdown in the endosperm was drastically reduced. When IAA, GA₃ and ABA were added together, the initial rate of modification in all seed treatments resembled that of the non-hormone treated seeds. The GA₃ induced stimulation of cell wall hydrolases between 3 and 4 days germination seen in the GA₃ and IAA + GA₃ treatments was, however, abolished in this case. The rate of modification in the later stages of germination was delayed in seeds with distal decapitation and nicks, while the terminal rate in whole seeds was identical to that of the control seeds.

Rootlet and Coleoptile Growth

As a measure of embryo growth, rootlet length and emerged coleoptile length were measured on seeds prior to sectioning for measurement of the endosperm modification. The results are presented in three figures: whole control seeds (Fig. 4); distally decapitated seeds (Fig. 5); nicked seeds (Fig. 6).

Neither distal decapitation nor nicking had any effect per se on rootlet or coleoptile growth. Due to a tendency of the coleoptile to emerge from the point of least resistance (i.e., the first

176

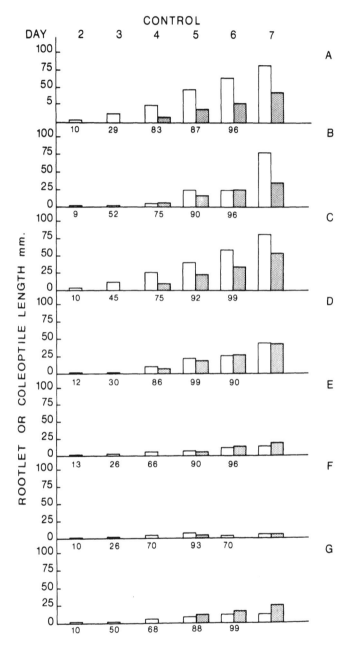

Figure 4.   Rootlet and Coleoptile growth during germi-
             nation. White columns - rootlet length;
             shaded columns - coleoptile length. For ex-
             planation of treatments see EXPERIMENTAL DE-
             SIGN.

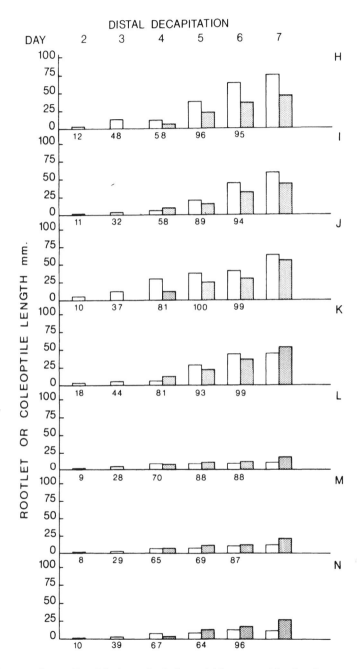

Figure 5.   Rootlet and Coleoptile growth during germi-
             nation of distally decapitated seeds. For
             explanation of treatments see EXPERIMENTAL
             DESIGN.

178

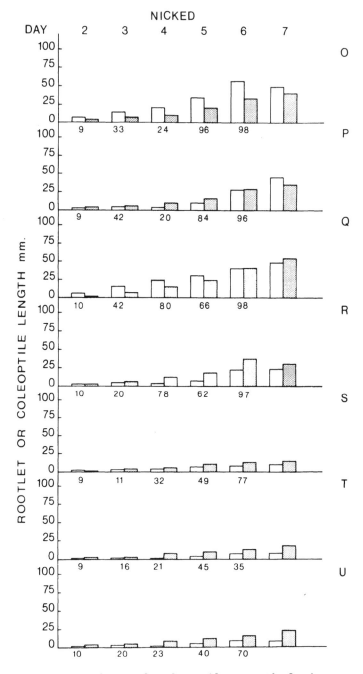

Figure 6.   Rootlet and Coleoptile growth during germi-
             nation of nicked seeds. For explanation of
             treatments see EXPERIMENTAL DESIGN.

nick), it was possible to measure coleoptile length earlier in nicked seeds.

IAA clearly inhibited rootlet formation for the first 4 to 5 days. The observed increase in rootlet length after the 5th day could reflect the fact that IAA was not renewed during the course of the experiment. IAA also exhibited a lesser inhibitory effect on coleoptile growth. There was no clear distinction between the different seed treatments with regard to this hormone.

GA_3 clearly affected coleoptile growth more than rootlet growth. Coleoptile growth was stimulated in all seed treatments. GA_3 has previously been shown to exert similar effects on coleoptiles, leaf sheaths and internodes of cereal plants (Thomas et al. 1973).

Combination of IAA and GA_3 totally changed the relative rates of elongation of rootlets and coleoptiles. Rootlet growth was inhibited as in the IAA treatment, and coleoptile growth stimulated as seen with GA_3. It is of interest to note that in this IAA + GA_3 treatment, there was no late recovery of rootlet growth as was seen with IAA treatment alone.

Addition of ABA strongly inhibited both rootlet and coleoptile growth. There was a clear independence of hormone effect on embryonic parameters and mechanical treatment of the seeds. ABA inhibited rootlet growth even more severely than coleoptile growth. The inhibitory effects on rootlets and coleoptiles of IAA + ABA together were even greater than those of ABA alone.

When GA_3 was used in combination with IAA and ABA rootlet growth was severely inhibited and the coleoptile growth reduced to between 35% to 50% of the control seeds. This inhibition of coleoptile growth was, however, not as high as that seen with ABA alone or with IAA and ABA in combination.

DISCUSSION

The results clearly demonstrate that extensive cell wall breakdown can occur in barley seeds treated with hormones which effectively inhibit growth of rootlets and coleoptiles (i.e., ABA and IAA). In such cases, the initial rate of cell wall breakdown is not affected significantly. The reduction in modification with ABA is first apparent after approximately 3 days. It has previously been shown that in the barley variety Nordal germinated at 15°C, the production and transport of both alpha-amylase and cell wall hydrolases during the first three days of germination appears to be independent of an active hydrolase synthesising system in the aleurone layer (Gibbons 1981).

It is true that in any work with phytohormones, the concentration of the hormone is of major importance. In the present experiments concentrations have been chosen which gave positive effects

corresponding to those previously published. That the sites of synthesis and release mechanisms of hydrolases, and the sites for rootlet growth and coleoptile growth may exhibit different concentration
responses for the applied phytohormones remains undeniable. The
present data, however, indicate that the initial stages of germination are under a different set of control factors than the later
stages, and that active embryo growth is not a prerequisite for the
initiation of internal modification of the grain.

As the internal modification of cereal grains is the practical
consequence, and thus the problem, of preharvest sprouting, the
results presented in this paper provide a basis for reassessment of
the validity of plant breeding programs and screening techniques in
this field.

REFERENCES

AAstrup, S., Gibbons, G.C., and Munck, L. 1981. A rapid method for
estimating the degree of modification in barley malt by measurement of cell wall breakdown. Carlsberg Res. Commun.
46:77-86.

Chrispeels, M.J., and Varner, J.E. 1967. Hormonal control of enzyme
synthesis: on the mode of action of gibberellic acid and abscisin in aleurone layers of barley. Plant Physiol. 42:1008-1016.

Gibbons, G.C. 1980a. Immunohistochemical determination of the
transport pathways of α-amylase in germinating barley seeds.
Cereal Res. Commun. 8:87-96.

Gibbons, B.C. 1980b. On the sequential determination of α-amylase
transport and cell wall breakdown in germinating seeds of
Hordeum vulgare. Carlsberg Res. Commun. 45:177-184.

Gibbons, G.C. 1981. On the relative role of the scutellum and
aleurone in the production of hydrolases during germination of
barley. Carlsberg Res. Commun. 46:215-225.

Mozer, T.J. 1980. Control of protein synthesis in barley aleurone
layers by the plant hormones gibberellic acid and abscisic
acid. Cell 20:479-485.

Munck, L., Gibbons, G.C., and Aastrup, S. 1981. Chemical and structural changes during malting. Proc. European Brewery Conven-
tion Congress, Copenhagen, 11-30.

Okamoto, K., Kitano, H., and Akazawa, T. 1980. Biosynthesis and
excretion of hydrolases in germinating cereal seeds. Plant &
Cell Physiol. 21:201-204.

Thomas, M., Ranson, S.L., and Richardson, J.A. 1973. Plant Physiology. 5th ed. Longman, London, pp 768-773.

Carbohydrates and Germination Inhibitors During Corn Seed Maturation and Germination

María Luisa Ortega-Delgado and
Ofelia Vega-Vásquez, *Colegio de Postgraduados, Chapingo, México*

SUMMARY

A field experiment was conducted in 1981 with two varieties of maize (*Zea mays* L.), H-139 and a local variety of San Antonio Tlacamilco, Pue. The objective was to study the changes in carbohydrates during several stages of corn seed development and also in early sprouting. In order to make comparisons, the carbohydrate content in germination of mature quiescent seeds was also determined. The presence of germination inhibitors in mature and immature seeds was observed. It was found that sucrose was the major soluble sugar in the endosperm and embryo of both varieties, but its content was reduced at 55 days after female flowering in the sprouted H-139 grain. At this stage the inhibitors of germination and seedling growth were reduced also. The inhibitory effects are similar to those observed by coumarin and abscisic acid. Maltose, as an indication of amylase activity, was found only in endosperm of germinated mature seeds.

Key words: *Zea mays*, early sprouting, starch, carbohydrates, germination inhibitors, seedling growth inhibitors.

INTRODUCTION

Early sprouting of corn was observed by Mangelsdorf (1923) . He isolated mutants characterized by seeds that germinate while still attached to the ear, before full maturity has been reached.

Reports of sprouting damage in maize were made by Routchenko and Soyer in France (1972). Their studies demonstrated that the problem had its origen in the low availability of Fe and Mn in the soil, together with genetic characteristics of the plant. González (1976) made similar observations in Venezuela. The damage was also observed in high valleys of Tlaxcala and Puebla in central México in 1976. (Félix, 1981).

In the present paper changes in carbohydrates and inhibitors of growth and germination have been studied in relation to grain development and early sprouting .

MATERIAL AND METHODS

Field experiments.- A trial was conducted with two varieties of maize plants: A double cross hybrid, H-139, and a local open pollinated variety of San Antonio Tlacamilco, Puebla. These varieties were planted on april 10, 1981 in a solid block with no replications. Rows were 80 cm apart and plants were 50 cm within the rows. The first sample corresponded to the immature embryo stage, 25 days after female flowering (d.a.f.), the second sample to the milky stage of grain (40 d.a.f.), and the third to the milky dough stage of grain development (55 d.a.f.).

Germination of mature seeds.- Seeds were germinated with distilled water on the dark at room temperature, for 3 to 7 days and thereafter soluble carbohydrates were determined.

Determination of soluble sugars.- Soluble sugars were extracted from the samples with 80% ethanol. One aliquot of the extract was analyzed in the Technicon carbohydrate automatic analyzer with a gradient of boric acid/sodium chloride (Kesler, 1967; Ortega and Rodríguez 1979).

Starch analysis.- The residue free of soluble sugars was hydrolized with 1% diastase followed by a second acid hydrolysis with hydrochloric acid (d=1.185) and quantified by the Nelson method (Nelson, 1944). Starch=glucose x 0.90.

Extraction and bioassay of inhibitors.- Two hundred whole corn seeds were extracted two times with 80% methanol. Extracts were combined and reduced to the aqueous fraction by rotary evaporation under reduced pressure at 35°C. The aqueous solution was lyophilized. From this material two extractions with ethyl ether were made, a basic extract at pH 8.2 and an acid extract at pH 2.8. These extracts were purified by thin layer chromatography eluted with isopropanol:ammonia: water (10:1:1 v/v). The region between Rf=0.3-0.9 was scraped off from the chromatograms and dissolved in 10 ml of distilled water; 50 lettuce seeds were spread over a glass Petri dish and sprayed with the extract. The dishes were covered and the seeds were allowed to germinate in the dark at room temperature for 4 days. Controls were prepared with 10 ml of distilled water. At the end of the assay period the percentage of germination in each dish was determined. The inhibitor content of extracts was compared with controls made with abscisic acid (ABA) and coumarin.

RESULTS

Field experiments. The H-139 variety showed immature sprouting (23%) at 55 d.a.f. and the local variety was almost not susceptible (5.4% of sprouting).

Soluble sugars.- Fig.1 shows the percentage of soluble sugars in endosperms and embryos of the local variety. In the endosperm from 25 to 40 d.a.f. there was a reduction

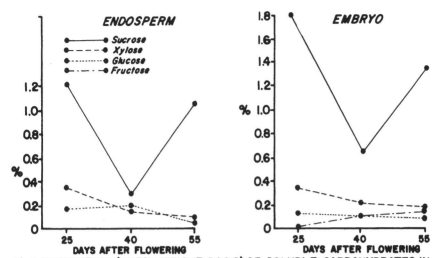

Fig.1. PERCENTAGE (FRESH WEIGHT BASIS) OF SOLUBLE CARBOHYDRATES IN
DEVELOPING ENDOSPERM AND EMBRYO OF LOCAL VARIETY.

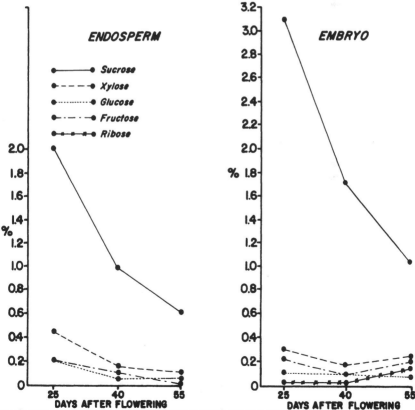

Fig.2. PERCENTAGE (FRESH WEIGHT BASIS) OF SOLUBLE CARBOHYDRATES IN
DEVELOPING ENDOSPERM AND EMBRYO OF H-139. THE SAMPLE AT 55
DAYS AFTER FLOWERING ALREADY HAD GERMINATED IN THE EAR.

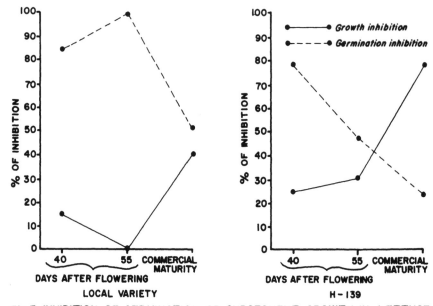

Fig.3. INHIBITION OF GERMINATION OR SUBSEQUENT GROWTH IN LETTUCE
SEEDS BY EXTRACTS OF WHOLE CORN SEEDS OF DIFFERENT STAGES
OF DEVELOPMENT NONE OF THE SEEDS OF EITHER VARIETY WERE
GERMINATING IN THE EAR.

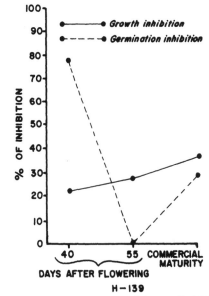

Fig. 4. INHIBITION OF GERMINATION OR SUBSEQUENT GROWTH IN LETTUCE
SEEDS BY EXTRACTS OF WHOLE CORN SEEDS (H-139) AT 40 AND 55
DAYS AFTER FLOWERING AND AT COMMERCIAL MATURITY. THE SAMPLE
AT 55 DAYS AFTER FLOWERING AND COMMERCIAL MATURITY CONSISTED
OF GERMINATED SEEDS.

of sucrose from 1.16% to 0.29%, and by 55 d.a.f. there
was an increase to 1.04%. The embryo had similar figures.
Xylose and glucose did not have a great quantitative
variation, but after 40 d.a.f., fructose could be detected in the embryo. By 40 d.a.f. sucrose had also de -
creased sharply in the endosperm of H-139, and continued
to decrease until 55 d.a.f.; at this stage there was
sprouting in the ear (Fig.2). The reduction of sucrose
was still more pronounced in the embryo of H-139. Ungermi
nated grains of H-139 belonging to the same ear had almost
similar results. Xylose, glucose and fructose in endo -
sperm of sprouted grains of H-139 had a tendency to reduction but in
the embryo ribose appeared in the sprouted seeds (Fig.2).

In order to make comparisons, mature seeds of the
local variety and H-139 were germinated in the dark and
analysed for soluble carbohydrates (tables 1 and 2). In
the local variety, besides the other sugars detected in
developing seeds, there were maltose and mannose in the
endosperm and small quantity of rhamnose in the embryo.
Maltose was also present in the endosperm of H-139.

TABLE 1.- Percentage of soluble sugars (fresh weight
basis) in germinated seeds. (seven days) of
the local variety.

S U G A R S	E M B R Y O %	E N D O S P E R M %
SUCROSE	1.86	0.69
MALTOSE	0.00	0.56
RHAMNOSE	0.03	0.00
RIBOSE	0.00	0.24
MANNOSE	0.00	0.17
FRUCTOSE	0.28	0.19
XYLOSE	0.39	0.35
GLUCOSE	0.07	0.62

TABLE 2.- Percentage of soluble sugars (fresh weight
basis in germinated seeds (three days) of H-139

S U G A R S	E M B R Y O %	E N D O S P E R M %
SUCROSE	1.23	0.63
MALTOSE	0.00	0.22
FRUCTOSE	0.20	0.00
XYLOSE	0.31	0.51
GLUCOSE	0.08	0.11

Starch.- During the seed development the content of
starch increased rapidly from 25 to 55 d.a.f. At harvest,
the seeds had the maximal value, 66% of starch for the
local variety and 67% for H-139 (table 3). The water
content decreased until the harvest date.

Inhibitors of germination and growth.- For prelim -
inary observations, two of the most important germination
inhibitors were chosen as type substances for comparison
with the bioassay of inhibitors: Abscisic acid (ABA),

Rf=0.69, and coumarin, Rf=0.70. ABA inhibited germination of the seeds. In the presence of 4 to 28 ppm of coumarin, lettuce seeds germinated but the growth of the seedling was inhibited, depending on the concentration;30 ppm of coumarin did not permit germination at all. The basic extract did not inhibit germination or growth of lettuce seeds,while the acid extracts were very effective. During the early stages of development in both varieties there was a strong inhibitory effect of the growth of the lettuce seedling (76-84%) (Figs. 3 and 4). Also at commercial maturity the local variety and the unsprouted H-139 had 88% and 100% of total inhibition.

TABLE 3.- Percentage of starch and moisture of several samples of developing and mature seeds of the local variety and H-139.

DAYS AFTER FLOWERING	STARCH		MOISTURE	
	LOCAL VARIETY %	H-139	LOCAL VARIETY %	H-139
25	6.9	6.7	75.38	77.63
40	29.10	29.93	50.34	47.12
55	42.65	53.61	41.37	40.02
		54.72*		
MATURE SEED				
	66.46	67.17	7.30	8.12

*Sprouted

DISCUSSION

The study of carbohydrates during the development of the corn grain and its early sprouting gives an insight about the initial substances available to the process. Sucrose was the major soluble sugar in endosperm and embryo. The reduction of sucrose from 25 to 40 d.a.f. probably is due to the utilization of this sugar for embryo maturation (Early, 1951). From 40 to 55 d.a.f., sucrose increased sharply in the local variety, but there was a reduction of this sugar in H-139, probably due to growth of the seedling in the sprouted grain.

During the normal development and early sprouting of the grain, xylose was always present, and ribose only appeared in the embryo during the early sprouting of H-139 grains. This means that the pentose phosphate pathway is operating in these stages of development (Gordon, 1979), as in the normal germination of mature grains (tables 1 and 2).

During the normal germination of mature seeds of the local and H-139 varieties, maltose was always present in the endosperm, probably as an indication of amylase activity (Kruger, 1976; Olered, 1976; Gordon, 1979). It is interesting that the inhibitory substances were reduced during the early sprouting of the H-139 variety(55 d.a.f.) Probably this is an indication that the absence of inhibitors, among other factors, could favor the growth of the seedling (King, 1976; Radley, 1976; Stoy and

and Sundin, 1976).

REFERENCES

EARLY, E.B., 1951. Percentage of carbohydrates in kernels of station reid yelow dent corn at several stages of development. Plant Physiol. 27:184-190.

FELIX, G.R., 1981. Germinación prematura del maíz en México. Tesis de Maestría en Ciencias (Fitopatología). Colegio de Postgraduados, Chapingo, México.

GONZALEZ, N.C., 1976. Germinación de granos inmaduros en mazorcas de maíz. Agronomía Tropical 26:359-362.

GORDON, I.L., 1980 Germinability, dormancy and grain development. Cereal Res. Commun. 8:115-129.

KESLER,R.B., 1967. Rapid quantitative anion-exchange chromatography. Anal Chem. 39:1416-1422.

KING, R.W., 1976. Abscisic acid in developing wheat grains and its relationship to grain growth and maturation. Planta 132:43-51.

KRUGER, J.E., 1976. Biochemistry of pre-harvest sprouting in cereals and practical applications in plant breeding. Cereal Res. Commun. 4:187-194.

MANGELSDORF, P.C., 1923. The inheritance of defective seeds in maize. J. Heredity 14:119-125.

NELSON, N., 1944. A photometric adaptation of the Somogyi method for the determination of glucose. J. Biol. Chem. 153:373.

OLERED, R., 1976. alfa-amylase isozymes in cereals and their influence of starch properties. Cereal Res. Commun. 4:195-199.

ORTEGA D., M.L. y Rodríguez C.C., 1979. Estudio de azúcarez solubles en semillas de frijol (Phaseolus vulgaris L.) Agrociencia 37:17-24.

RADLEY, M., 1976. The development of wheat grain in relation to endogenous growth substances, J. Exp.Bot.27: 1009-1021.

ROUTCHENKO, W. et Soyer, J.P., 1972. Causes de la germination sur plante de grains inmatures de máis. Donnés complementaires. Ann. Agron. 23:445-459.

STOY, V. and Sundin, K., 1976. Effects of growth regulation substances in cereal seed germination. Cereal Res. Commun. 4:157-163.

Fluridone Induction of Vivipary During Maize Seed Development

Franklin Fong, Don E. Koehler and
James D. Smith, *Texas A&M University,
College Station, Texas, U.S.A.*

INTRODUCTION

Vivipary in corn seeds was first described by Lindstrom (1923) and Mangelsdorf (1923). Robertson (1955,1975) described the pleiotropic effects of many separately mapping genes which caused vivipary in maize. Smith et al. (1978) concluded that at least two distinct classes of viviparous genes could be identified: type I which includes genes such as vp which have normal abscisic acid (ABA) content and are not sensitive to ABA; and type II includes vp-2, vp-9, which have reduced carotenoid content. Robichaud et al. (1980) have recently added vp-5 to Class II, and proposed that vp-7, vp-8 and vp-9 should be in a separate class as these appear to have more complex alterations.

The carotenoid content of whole seed homozygous for viviparous genes at 19-25 days after pollination (DAP) have been examined by Fong et al. (1981). Except for vp there is a very specific block in carotenoid biosynthesis in these seeds as summarized in Fig. 1.

All the class II mutants show a block at some step before zeaxanthin. The class I mutant, vp, shows a typical wild type accumulation of zeaxanthin.

These results on carotenoid content and viviparous gene expression leads to the conclusion that there is a causal relationship between absence of xanthophylls and appearance of vivipary. This hypothesis was tested by examining the effectiveness of a carotenoid biosynthesis inhibitor on the expression of vivipary.

Fluridone: 1-methyl-3-phenyl-5 (3-trifluoromethyl phenyl) -4(1H) -pyridinone

CAROTENOID INTERMEDIATES:

```
      phytoene
----- ↓ ---- vp-2*, vp-5, vp-x₁
      phytofluene
----- ↓ ---- w-3
      ζ-carotene
----- ↓ ---- vp-9
      neurosporene
            ↓
      δ-, γ-carotene
----- ↓ ---- ps( = vp-7)
      a-, β-carotene
----- ↓ ---- al( = y-3)
      a-, β-cryptoxanthin
----- ↓ ---- y-9
      zeaxanthin, lutein,
                  wild type, vp
```

(*leaky mutant, crypotoxanthin present)

Fig. 1.   Biosynthesis of carotenoids.

MATERIALS AND METHODS

The yellow dent corn inbred Tx5855 and Va35 were planted at the Texas A&M Brazos River Research Farm near College Station. Normal cultural practices for the area were followed throughout the growing season. Developing ear shoots were bagged prior to silking. Approximately 150 ears of Va35 and 100 ears of Tx5855 were sibling pollinated. Treatments with fluridone were applied to 7 ears on alternate days beginning 3 DAP and ending 23 DAP for Va35 and ending 19 DAP for Tx5855.

At the appropriate time husks were pulled back, silk removed, and the upper half of the ear covered with a glassine bag before spraying the lower half with 100 µg/mL fluridone in 1% acetone. After the excess solution drained, the bag was removed, husks refolded over the ear, and the shoot was re-bagged and labelled. Control ears were sprayed with 1% acetone and gave no response to parameters being measured as compared with untreated ears. Five ears of each age group were harvested at 25 to 27 DAP, husked, sealed in plastic bags and stored at -50°C until analyzed. Two ears of each age group were harvested at 45-50 DAP or at full maturity, and stored at room temperature.

Carotenoids were extracted from fluridone-treated or untreated seeds or each ear at 25 to 27 DAP. About 75-100 seeds were picked, weighed and ground in acetone with 1 g $CaCO_3$ and sand. The material was re-extracted at least 5 times with fresh acetone over a 24-36 hr period. Procedures involving saponification and preparation of

190

extracts were as described by Vaisberg and Schiff (1976) and Jensen and Jensen (1971). The carotenoids were separated on silica gel TLC using 20% (v/v) ethyl acetate in methylene chloride (Vaisberg and Schiff 1976). Phytoene and phytofluene, UV-absorbing carotene intermediates, are easily separated from other UV absorbing materials which remain at the origin. The location of these compounds on silica gel was ascertained by viewing under UV light. Bands were scrapped from the plate, eluted with ether, resuspended in a known volume of petroleum ether and quantified from its absorption spectrum, using appropriate extinction coefficients (Goodwin 1952).

RESULTS

Seeds were treated once with fluridone. Seeds harvested at 25 DAP with no fluridone treatment (A), treatment at 19 DAP (B), at 9 DAP (C), at 7 DAP (D), and at 3 DAP (E) are shown in Fig. 2. The basal (left) portion of C,D and E are white, while the apical portions are yellow. Ears A and B are uniformly yellow. Vivipary is most obvious in C and all the ears treated between 9 and 13 DAP. Coleoptiles emerging from the seed are shown in Fig. 3.

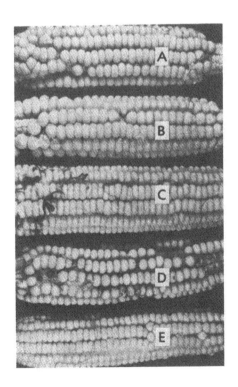

Fig. 2. Effect of fluridone treatment on seeds harvested at 25 DAP: no treatment (A), treatment at 19 (B), 9 (C), 7 (D) and 3 (E) DAP.

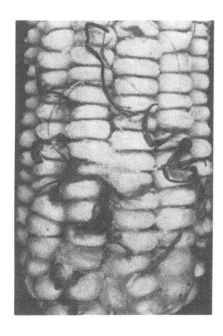

Fig. 3. Emergence of coleoptiles in fluridone-treated
seed at 9 DAP (C) in Fig. 2.

The absorption spectrum of carotenoid extracts from fluridone-
treated seeds (Fig. 2) are shown in Fig. 4. The major peak of

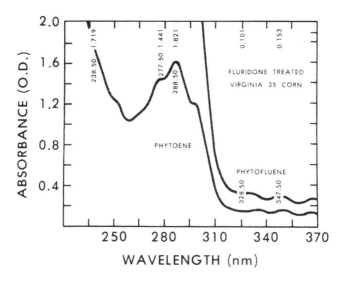

Fig. 4. Absorption spectrum of carotenoid extracts from fluridone-
treated seeds.

phytoene absorption is in the UV region at 285 nm. Minor peaks of phytofluene occur at 387 nm. Phytoene is the expected accumulation product since fluridone inhibits the desaturation of phytoene (Bartells and Watson 1978). In ears which were harvested at 40 to 45 DAP, only those which were treated between 9 and 13 DAP were viviparous in nearly every seed.

Seeds harvested at 45 DAP are shown in Fig. 5. All seeds are shown with germ side upwards and labelled with the age when treated with fluridone. Control seeds and seeds treated at 19 DAP have a completely white colored germ region. All other seeds show a dark brown band of tissues. The dark region is very small in those seeds treated at young ages, and appears to develop into the emerging embryo of viviparous seeds.

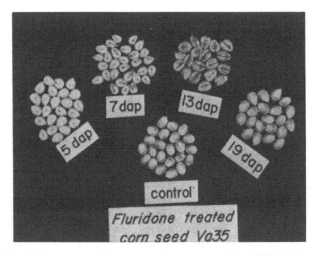

Fig. 5. Fluridone-treated corn seeds at maturity (45 DAP).

Control seeds taken from apical or the untreated half of each treated ear and seeds treated with fluridone 15 DAP produced normal green seedlings.

No germination occurred in other fluridone-treated seed except for an occasional root which formed.

The phytoene content of fluridone-treated seeds are shown in Fig. 6. All seeds were 25 to 27 DAP at harvest. The control values are from seeds taken from the apical portion of each ear. Although seeds were 25 to 27 DAP in age, control seeds show higher phytoene content in ears treated with fluridone between 3 and 9 DAP compared with ears treated between 11 and 17 DAP. Despite these differences, there are large differences in phytoene content of fluridone-treated seeds at certain ages. Seeds treated at 3 or 9 DAP have 32-38% less phytoene than controls. Yet seeds treated at 7, 15 or 17 DAP have 259%, 220% or 560%, increases, respectively, in phytoene content compared with control values.

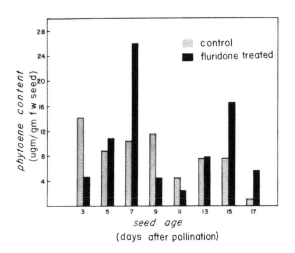

Fig. 6. Phytoene content of fluridone-treated seeds (25 to 27 DAP).

Results of fluridone treatment may be summarized as follows:

Seed: Age (DAP)	3	5	7	9	11	13	15	17	19	21	23
Color	−	−	−	−	−	−	−/+	+	+	+	+
Viability	−	−	−	−	−	−	+	+	+	+	+
Vivipary	−	−	−	+	+	+	−	−	−	−	−
Phytoene	−	nc	+	−	nc	nc	+	+	..	..	..

nc: no change

DISCUSSION

Vivipary in maize is probably controlled by at least two different mechanisms: one related to embryo responsiveness and another related to ABA content (Smith et al. 1978; Robichaud et al. 1980). A reduction in seed color is also evident in mutants of the second group (Robertson 1975; Fong et al. 1981) e.g. vp-2, vp-5, vp-7, vp-9, w-3, y-3, y-9. Chemical inhibitors of carotenoid bio-synthesis alter seed color (Fig. 2) and induce vivipary (Fig. 3).

Seed color may be altered by applying fluridone between 3 and 15 DAP. Phytoene is the predominant carotenoid accumulated in fluridone-treated seeds. A 2 to 5 fold increase in phytoene results when seeds are treated at 7 or 15 DAP. Smaller accumulation of phytoene occurs when seeds are treated at other days. A 70% reduction in phytoene content only occurred when seeds were treated at 3 DAP. These large fluctuations in phytoene content suggest that carotenoid metabolism is very active in at least three different periods of seed maturation: between 3 to 5 DAP, at 7 DAP and at 15 DAP. One possibility is that these periods of activity may represent metabolic differences between endosperm and/or embryo tissues.

Seed viability and dormancy are also only affected by fluridone at certain ages of seed maturation. The induction of dormancy is inhibited by fluridone only if the seeds are treated between 9 and 13 DAP. The apparent dormancy that occurs with seeds treated at younger ages than 9 DAP is probably caused by embryo death during development. Such seeds are not viable and thus only appear dormant. Fatty acid desaturation reactions may be inhibited by fluridone (St. John 1976) and this may be the basis of its effects on seed viability.

Fluridone-induced changes in phytoene content, viability, and vivipary occur only at specific seed ages. Our working hypothesis for these fluridone effects is as follows:

Phytoene⫿⫿⟩Phytofluene⫿⟩⫿⟩⫿⟩⫿⟩xanthophylls⟩⟩ABA⫿⟩Dormancy

Fluridone inhibition: ⫿

Genetic mutation: ⫿ Type I   ⫿ Type II

During the period from 9 and 13 DAP, the embryo and/or endosperm are committed to dormancy depending on ABA content or ability to respond to ABA. Any disruption of ABA production such as loss of key enzymes for production of ABA precursors, carotenoid biosynthesis, or chemical inhibition of biochemical pathways would result in the absence of dormancy and continued viviparous development of embryos.

Carotenoids may be the precursors of ABA based on the stereochemical similarities of ring structures of carotenes and ABA (Milborrow 1972) and on the photoconversion and chemical conversion of violaxanthin to xanthoxin, an ABA-like hormone found in many plants (Burden and Taylor 1976). Also consistent with our hypothesis is the observation that fluridone-treated corn seedlings are hypersensitive to gibberellic acid (Devlin et al. 1980), the antagonistic hormone of ABA. Walton (1980) recently concluded that there is no clear evidence to discount the possibility that 'C-40' compounds may be ABA precursors.

At present we do not know what anatomical or histological changes are occurring in 9 to 13 DAP seeds. Since temperature influences seed maturation, similar chronological ages in another climate may or may not correspond to developmental changes (Lampe 1931). Thus, it is with great caution that we would extrapolate our findings to seeds grown in a different climate.

Using a carotenoid inhibitor such as fluridone to probe seed development provides physiologists and biochemists with a new experimental system to study premature sprouting. Not only does it provide a large quantity of plant material for study, but it also raises some very interesting new questions about our understanding of carotenoids and ABA, about unsaturated compounds and seed viability, and about the regulation of carotenogenesis in corn seeds.

CONCLUSIONS

Abscisic acid plays an important role in controlling seed development. Without ABA or when the embryo lacks the ability to respond to ABA, the developing embryo continues to grow and germinate prematurely. Fluridone, a chemical inhibitor of carotenoid biosynthesis, effectively inhibits dormancy if applied to corn seeds between 9 and 13 days after pollination. Thus, premature sprouting or vivipary occurs when there is inhibition of carotenoid biosynthesis caused by either genetic mutation or by chemical inhibition. The cumulative evidence from our studies on vivipary and carotenoids in corn seeds strongly support the bioconversion of carotenoids to ABA.

REFERENCES

Bartells, P.G., and Watson, C.W. 1978. Inhibition of carotenoid synthesis by fluridone and norflurazon. Weed Sci. 26:198-203.

Burden, R.S., and Taylor, H.J. 1976. Xanthoxin and abscisic acid. Pure Appl. Chem. 47(2/3)203-209.

Devlin, R.M., Kisiel, M.J., and Kostusiak, A. 1980. Enhancement of gibberellic acid sensitivity in corn (Zea mays) by fluridone and R-40244. Weed Sci. 28:11-12.

Fong, F., Koehler, D.E., and Smith, J.D. 1981. Genetic and chemical control of carotenogenesis and vivipary in maize seed. Plant Physiol. Suppl. 67:142.

Goodwin, T.W. 1952. Carotenogenesis. III. Identification of the minor polyene components of the fungus Phycomyces blakesleeanus and a study of their synthesis under various cultural conditions. Biochem. J. 50:550-558.

Lampe, L. 1931. A microchemical and morphological study of the developing endosperm of maize. Bot. Gaz. 91:337-377.

Liaaen-Jensen, S., and Jensen, A. 1971. Quantitative determination of carotenoids in photosynthetic tissues. Methods Enzymol. 23:586-602.

Lindstrom, E.W. 1923. Heritable characteristics of maize. XIII. Endosperm defects — sweet defective and flint defective. J. Hered. 14:127-135.

Mangelsdorf, P.C. 1923. The inheritance of defective seeds in maize. J. Hered. 14:119-226.

Milborrow, B.V. 1972. Stereochemical aspects of the formation of double bonds in abscisic acid. Biochem. J. 128:1135-1146.

Robertson, D.S. 1955. The genetics of vivipary in maize. Genetics 40:745-760.

Robertson, D.S. 1975. Survey of the albino and white endosperm mutants of maize their phenotypes and gene symbols. J. Hered. 66:67-74.

Robichaud, C.S., Wong, J., and Sussex, I.M. 1980. Control of in vitro growth of viviparous embryo mutants of maize by abscisic acid. Developmental Genetics 1:325-330.

St. John, J.B. 1976. Manipulation of galactolipid fatty acid composition with substituted pyridazinones. Plant Physiol. 57:38-40.

Smith, J.D., McDaniel, S., and Lively, S. 1978. Regulation of embryo growth by abscisic acid in vitro. Maize Genetics Coop Newsletter 52:107-108.

Vaisberg, A.J., and Schiff, J.A. 1976. Events surrounding the early development of Euglena chloroplasts. Plant Physiol. 57:260-269.

Walton, D.C. 1980. Biochemistry and physiology of abscisic acid. Ann. Rev. Plant Physiol. 31:453-89.

Role of Glucose-6-Phosphate Dehydrogenase in Corn Seed Germination

Estela Sánchez de Jiménez and Jesús Quiroz,
Universidad Nacional Autónoma de México, México

SUMMARY

Glucose catabolism through the PPP was evaluated in corn seed germination by in vitro measurements of G6PDH activity in axis and scutellum, and $^{14}CO_2$ evolution from these tissues when incubated either with ^{14}C-1 or ^{14}C-6 glucose. The results indicate there is a significant increase in G6PDH and $^{14}CO_2$ from ^{14}C-1 glucose in the first 6 hs of seed imbibition. An G6PDH inhibitor was detected in axis of quiescent seeds. Partial purification and characterization of this inhibitor indicates that it is a protein of approximately 13,000 daltons M.W.

The above biochemical parameters were used in a field experiment designed to investigate on the nature of corn pre-harvest sprouting. The results show that the activity of both enzyme and inhibitor are altered on these varieties.

INTRODUCTION

The degradation of glucose through the pentose phosphate pathway (PPP) provides the cell with a) extramitochondrial reducing power for NADP and b) pentoses for nucleic acid synthesis, both of which are required for seed germination.

It has been suggested that PPP plays an important regulatory role for germination of certain gramineous seeds (Roberts and Smith 1977). An increase of glucose catabolism through this pathway has been demonstrated in relation to hazel (Gosling and Ross 1980) and oat seed dormancy release (Kovacs and Simpson 1976). A sharp increase in glucose-6-phosphate dehydrogenase (G6PDH, E.C.1.1.1.49), the first enzyme of this pathway, has also been described for germination of non-dormant oat seeds (Upadhyaya et al. 1981).

Very little is known about the physiological and biochemical causes of the pre-harvest sprouting of corn, (Routchenko et Soyer 1972, Felix 1981). This study explores the G6PDH activity as: a) a marker to detect the onset of corn germination, and b) a screening method for susceptibility to pre-harvest sprouting.

MATERIAL AND METHODS

Biological Material.- Dry seeds from several corn varieties: hybrids H-30 and H-139, an improved variety VS-22,
and two local varieties from Puebla and Tlaxcala (Tlacamilco and Zaragoza, respectively), were used for this study. The seeds were desinfected, germinated in sterile distilled water at 25°C, and then the embryonary axis and the
scutellum isolated. In the respirometric experiments, previously disected tissues were incubated and processed.

A field experiment was conducted with two varieties
of corn plants: the hybrid H-139, and the local variety of
Sn. Antonio Tlacamilco, Puebla (Tlacamilco). Five samples
of ears were collected every 15 days along the grain maturation period, from the immature embryo stage, 25 days a-
fter female flowering (d.a.f.), to the stage of physiological maturation of the grain (85 d.a.f.). Ten ears were
collected at each stage of maturation and the seeds from
the 2/3 central part of the ears collected and pooled. Visibly detected sprouted seeds were discarded.

Assay of Glucose-6-Phosphate Dehydrogenase.- Either
axis or scutellum were ground at 0°C in 0.1 M glycine buffer, pH 7.5 centrifuged at 10,000 xg 30 min and then at
120,000 xg 60 min. G6PDH activity was determined in the
supernatant following NADP reduction as a function of time
at 340 nm at 30°C. The enzyme assay contained: 15 mM glycine buffer pH 7.5, 10 mM $MgCl_2$, 2 mM G-6-P (K salt), 12
μM NADP, 60 μl enzyme extract in a final volum of 3.0 ml.
Endogenous reduction was substracted in each sample. One
enzyme unit corresponds to a change in absorbancy of 0.001
Units/min.

Inhibitor purification.- The postribosomal homogenates of the axis were frozen and thawed three times in order to destroy G6PDH activity, centrifuged and precipita-
ted with sulphate at 70% saturation. The precipitate was
dialized and filtered through a Sephadex G-200 colum
(2 x 86 cm). Fractions with G6PDH inhibiting activity were
pooled, concentrated and analysed by sucrose gradient centrifugation.

RESULTS

I. Normal Germination.

The pattern of G6PDH activity in embryonic tissues at
different times after imbibition was determined (fig. 1).
An increase in enzyme specific activity during the first
six hours of incubation was consistenly observed in the
axis and scutellum tissues of VS-22 and other non-sprout-
ing varieties (see later). Although the activity in the
scutellum is higher than in the axis, the change in activity during this initial period is more significant in the
later, since it more than doubles its original value after
6 hs (fig. 1). The presence of an inhibitor in the axis
extracts was detected. The inhibitory activity was followed in axis homogenates during the initial hours of germination and shown to decrease sharply during the first 6

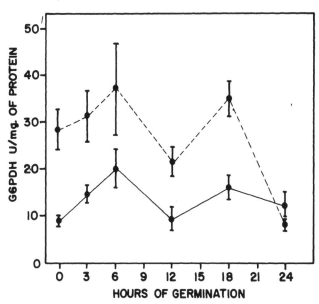

Fig.I. G6PDH ACTIVITY IN CORN TISSUES DURING GERMINATION.
Embryonary axis ——— , scutellum ----. Each value represents
average of five independant determinations. Vertical barrs = S.D.

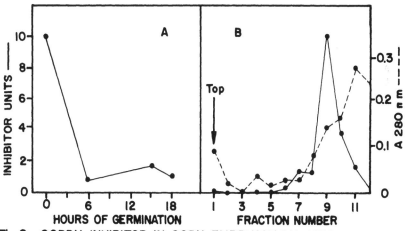

Fig. 2. G6PDH INHIBITOR IN CORN EMBRYONARY AXIS.
A- 0.I ml homogenate sample was tested against a fixed amount of
G6PDH activity. Relative inhibitor units were set for our assay
system.
B-5-20% sucrose gradient centrifugation analysis of the G6PDH
inhibitor was carried out at 38000 rpm, SW40 rotor in a L75
Beckman ultracentrifuge for 18 hs at 0° C. Inhibitor activity was
tested in each fraction.

hs (fig. 2a). The inhibitor activity was sensitive to pro-
teolisis and heat and was not dializable. It was partially
purified by sulfate precipitation and fractionation through Sephadex G-200. The inhibitory activity was found associated to a protein of about 13,000 daltons as show by
sucrose gradient centrifugation. The inhibitor was very
effective against G6PDH from corn axis and human red cell
lysate (fig. 2b), but only partially reduced the enzyme
activity from the scutellum.
    Axis, scutellum or whole embryos were incubated in a
Warburg respirometer at 25°C. Either C-1 or C-6 labelled
^{14}C-glucose was added at 0 or 5 hs incubation for 1-h pul-
se period and the $^{14}CO_2$ evolved was trapped on a filter
paper saturated with 30% KOH placed on the central vessel
of the incubation flask.

TABLE 1
RESPIROMETRIC EXPERIMENTS IN CORN EMBRYONIC TISSUES

Incubation Period*	$^{14}CO_2$ evolved (cpm/g fresh tissue)		
	Embryo	Scutellum	Axis
^{14}C-1 glucose			
0-1 h	4844	5405	37840
5-6 h	5263	6971	50685
^{14}C-6 glucose			
0-1 h	2278	2207	23476
5-6 h	2071	2388	14312
C-6/C-1 Ratio			
0-1 h	0.47	0.41	0.62
5-6 h	0.39	0.34	0.28

* Incubation system: 50 mM Tris-HCl buffer, pH 7.5, 10 mM
$MgCl_2$, 50 mM KCl, 2% sucrose and 3 μg of ^{14}C-glucose (61
mCi/mM) in a final volum of 1.0 ml. Values are average of
3 to 5 independant determinations.
    When ^{14}C-1 glucose was tested, an increase in the
amount of absorved radioactive CO_2 was observed during the
0-1 to 5-6 hs period, but not when ^{14}C-6 glucose was used
(Table 1). This increment was more noticeable in the axis
than in the scutellum and paralels the increase of G6PDH
activity observed in vitro during the 0-6 h period. Conse-
cuently, the C-6/C-1 ratio of $^{14}CO_2$ changes between the
0-1 to the 5-6 h-period, being again the largest change in
the axis (0.62 to 0.28, Table 1). This indicate an increase of glucose catabolism through the PPP during this period (Kanamori et al. 1979). These experiments suggest
that the G6PDH enzyme from embryonic axis might be regulated by a control, which switches the utilization of glucose from the glycolitic pathway to the PPP on the onset of
corn seed germination.
II. Pre-harvest Sprouting
    In an effort to approach the study of the pre-harvest
sprouting in an interdisciplinary manner, a field experiment was set in an area where strong corn pre-harvest
sprouting has been detected (Felix 1981).
    G6PDH activity was measured on the axis and the scu-
tellum from seed samples at various stages of maturation,

in two of the varieties planted in this experiment: the hybrid 139 and Tlacamilco. The activity of the scutellum did not show important fluctuations along the period of maturation (fig. 3). In contrast, the enzyme from the embryonic axis changed dramatically along this period. At the earliest stage (25 d.a.f.), when the embryo is still immature, no activity was detectable in the embryonic axis but, as maturation proceeded, the enzyme activity increased up to maximum value at the milky dough stage of grain development (55 d.a.f.). At this stage, a burst of sprouting was observed in both varieties. It is important to point out that seeds at the earlest stage, where no G6PDH was detectable, were incapable of germinating when placed in a favorable environment. However, at all other maturation stages seeds germinated under the same conditions.

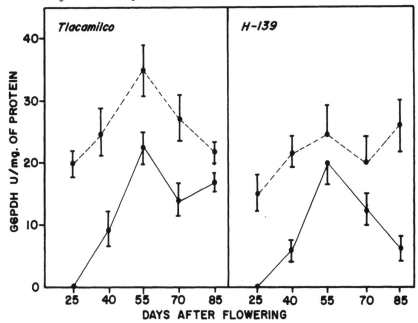

Fig. 3. G6PDH ACTIVITY IN CORN TISSUES DURING GRAIN MATURATION.
Embryonary axis ———, scutellum – – – –. Each value represents average of five independant determinations. Vertical barrs = S. D.

Screening for G6PDH activity was carried out among dry seeds of different corn varieties and different harvests with and without sprouting problem. The two varieties used in the previous experiment were also included (Table 2). The results showed that the levels of enzyme in the axis were lower than in the scutellum in the seeds which do not present the sprouting problem; however, the axis activity was enhanced in varieties where the phenomenon has been detected. A scutellum/axis (S/A) value lower than three was correlated with pre-harvest sprouting. It

is interesting to note that corn var. Zaragoza showed the
lowest S/A ratio and is one of the varieties most susceptible to early sprouting.

TABLE 2
G-6-P DEHYDROGENASE IN TISSUES OF CORN EMBRYOS

Corn Variety	Harvest	G-6-P-DH (Units/mg protein)		S/A Ratio	Sprouting
		Axis	Scutellum		
Tlaxcala	1977	8.0	30.0	3.3	-
VS-22	1979	9.8	28.0	3.3	-
H-30	1978	9.0	27.5	3.3	-
H-30	1980	14.0	44.0	3.0	-
H-30✦	1980	17.5	37.0	2.1	+
Tlacamilco✦	1980	28.0	35.0	1.2	+
Tlacamilco✦	1981	27.6	36.1	1.3	+
H-139✦	1981	19.3	44.6	2.3	+
Zaragoza	1977	26.0	25.0	0.9	+

✦ Harvested at Sn. Antonio Tlacamilco.

Finally, the G6PDH increase observed between 0-6 hr
(fig. 1) as well as the G6PDH inhibitor of the embryonary
axis were investigated in these seeds. It was found that
the seeds from varieties where pre-harvest sprouting has
been observed did not show G6PDH increase at initiate germination, nor was any detectable activity of G6PDH inhibitor present in their axis (Table 3).

TABLE 3
PERCENTAGES OF G6PDH ACTIVITY AT 6 HS OF SEED IMBIBITION

Corn Variety	Axis	Scutellum	Inhibitor
	Percentage		
VS-22✦	234	136	+
H-30✦	250	163	+
H-139	120	91	ND
Tlacamilco	67	167	ND
Zaragoza	47	60	ND

The activity found in dry seed tissues= 100%.
✦ Harvested in Chapingo. ND-Non detectable.

DISCUSSION

Large fluctuations of G6PDH activity were observed in
the embryonary axis along grain maturation and seed germination. Furthermore, a G6PDH inhibitor in the axis from
quiescent seeds was found, which almost disappears once
the germination process has started. Since G6PDH is the
key enzyme in the regulation of the PPP (Ashira and Koma-
meni 1976, Adkins et al. 1980), a regulatory mechanism for
this pathway seems to operate in corn embryonary axis in
relation to the biochemical onset of corn germination; this
suggestion is consistant with observations in other cereals
(Gordon 1980). The data regarding the pre-harvest sprouting of corn permit us to speculate that a repressive control for PPP could be set up in the embryonary axis, some
time along the period of grain maturation. If it fails,
some glucose shall be processed through this pathway since
G6PDH has pruved to be active even under limited hydration
(Stevens and Stevens 1979). This process would cause NADPH
to accumulate and be readily available for many other es-

sential reactions for germination to occur. The triggering of such reaction would committ the corn embryo to extemporaneous germination. This hypothesis must however, await the support of further experimental evidence.

ACKNOWLEDGEMENT

The present study is part of the Program for Academic Collaboration between the Universidad Nacional Autónoma de México-Colegio de Postgraduados, Chapingo and was supported by CONACYT (National Council for Science and Technology), grant No. PCDFNAL 790382.

REFERENCES

Adkins, S.W., Gosling, P.G. and Ross, J.D. 1980. Glucose-6-phosphate dehydrogenase and 6-phosphogluconic acid dehydrogenase of wild oat seeds. Phytochem. 19:2523-2525.

Ashihara, H. and Komamine, A. 1976. Characterization and regulatory properties of glucose-6-phosphate dehydrogenase from black gram (Phaseolus mungo). Physiol. Plant. 36: 52-59.

Felix, G.R. 1981. Germinación prematura del maíz en México. Tesis de Maestría en Ciencias (Fitopatología). Colegio de Postgraduados, Chapingo, Méx.

Gordon, I.L. 1980. Germinability, dormancy and grain development. Cereal Res. Commun. 8, 115-129.

Gosling, P.G. and Ross, J.D. 1980. Pentose phosphate metabolism during dormancy breakage in Corylus avellana L. Planta. 148: 362-366.

Kanamori, I., Ashihara, H. and Komamine, A. 1979. Changes in the activities of the pentose phosphate pathway and pyrimidine nucleotide biosynthesis during the growth of Vinca rosea cells in suspension culture. Z. Pflanzenphysiol. 93: 437-448.

Kovacs, M.I.P. and Simpson, G.M. 1976. Dormancy and enzyme levels in seeds of wild oats. Phytochem. 15: 455-458.

Roberts, E.H. and Smith, K.D. 1977. Dormancy and the pentose phosphate pathway. In: The Physiology and Biochemistry of Seed Dormancy and Germination. Khan. A. A. ed. Amsterdam Elsevier, pp. 388-391.

Routchenko, W. et Soyer, J.P. 1972. Causes de la germination sur plante de grains inmatures de máiz. Donnés complementaires. Ann. Agron. 23: 445-459.

Stevens, E. and Stevens, L. 1979. The effect of restricted hydration on the rate of reaction of glucose-6-phosphate dehydrogenase, phosphoglucose isomerase, hexokinase and fumarase. Biochem. J. 179: 161-167.

Upadhyaya, M.K., Simpson, G.M. and Naylor, J.M. 1981. Levels of glucose-6-phosphate and 6-phosphogluconate dehydrogenases in the embryos and endosperms of some dormant lines of Avena fatua during germination. Can. J. Bot. 59: 1640-1646.

Hydrolysis of Endosperm Protein in *Zea mays* (W64A × W182E)

A. Oaks, M. J. Winspear and S. Misra,
McMaster University, Hamilton, Ontario, Canada

INTRODUCTION

Metabolic processes involved in seed development and germination are in many respects symmetrical. In the former process storage components (starch, protein and lipid) are synthesized; in the latter those stored products are hydrolyzed and transported to the embryo. In either case metabolic events occurring in the endosperm and embryo are interrelated. Normally reactions leading to germination or to the hydrolysis of storage reserves are suppressed during the latter stages of seed development.

The initiation of germination, an embryo response, may be precocious or delayed. Extreme examples would be viviparous mutants in maize which germinate on the cob (McDaniel et al. 1977) and dormancy in wild oats (Avena fatua). With "normal" genotypes the onset of the ability to germinate can be altered by environmental conditions as indicated by variable responses of the induction of enzymes associated with germination or post-germination events from one crop year to the next (Goldbach and Michel 1976, Sawhney and Naylor, 1979). Two plant growth substances gibberellic acid (GA) and abscisic acid (ABA) appear to play a significant role both in permitting or preventing germination and in regulating the hydrolysis of storage products in the endosperm. The relative levels of these two hormones change with seed development and may be influenced by the environment (Walton 1980/81; Goldbach and Michel 1976). I would like to stress here that the control of sprouting or normal germination rests initially with metabolic events in the embryo, subsequently and, in response to signals from the embryo, digestion of the endosperm commences (see for example Chen and Chang 1972). Continued growth of the seedling after germination depends very much on the efficient hydrolysis of stored starch, protein, and lipid initially in the embryo itself, subsequently in the scutellum and finally in the endosperm (Toole 1924).

Prolamines (zein in maize) which are located in protein bodies, are rich in proline, leucine and glutamine, and the glutelins which form a structural support of protein bodies and amyloplasts are the major proteins in the cereal endosperm (Wall and Paulis 1978). These two protein classes are essentially water insoluble. Nevertheless they are hydrolyzed and they serve as the main source of reduced nitrogen available to the young seedling (Folkes and Yemm 1958; Oaks and Beevers 1964, Oaks, 1965, Sodek and Wilson 1973). Several workers have shown that while the endosperm is supplying nitrogen to the embryo that the use of exogenous nitrate is minimal (Srivastava et al, 1970; Blohm 1972). The enzymes involved in the hydrolysis of the endosperm proteins are probably both endopeptidase(s) and exopeptidase(s). Feller and Hageman (1978) have shown an increase in both endopeptidase(s) and carboxylpeptidase(s) in endosperm tissues after germination and a decrease in amino-peptidase activity. However it should be noted that there are likely to be many enzymes in each of the general classes of peptide hydrolase (Mikola 1982).

In this paper, I would like to consider the hormonal involvement in the regulation of the hydrolysis of storage proteins in the maize endosperm. Then I want to consider the processes involved in the initial solubilization of those storage proteins.

METHODS

Seed material (W64A × W182E) was supplied by the Wisconsin Seed Foundation, Madison WI. (USA) and was grown in a mixture of sand and vermiculite in a growth chamber (Conviron) at 28°C and a 16 hr. day (~ 150μ E/m^2/sec) or for extended periods in walk-in growth rooms under essentially similar conditions. Endosperm enzyme preparations and assays were carried out as described by Harvey and Oaks (1974a) and Feller and Hageman (1978). Commercial carboxypeptidase (Calbiochem.) and pronase (Cabiochem.) were prepared as described by Winkler and Schon (1979). The endosperm powder used for these digestions and those performed by the corn peptide hydrolases, was prepared from endosperms collected 30 days after pollination. They were lyophilized, ground to a powder in a Moulinex coffee mill and incubated for 48 hr. with the appropriate enzyme additions. The nitrogen assays (α-NH$_2$N or total N) were performed essentially as described previously (Harvey and Oaks 1974a,b).

RESULTS AND DISCUSSION

1. <u>Induction of the Hydrolase Enzymes</u>: We have shown that there is an increase in endopeptidase activity in the maize endosperm which parallels the loss of endosperm protein. A similar loss of endosperm protein

and appearance of endopeptidase activity occurs in de-
embryonated endosperms (Fig. 1). The endopeptidase(s)

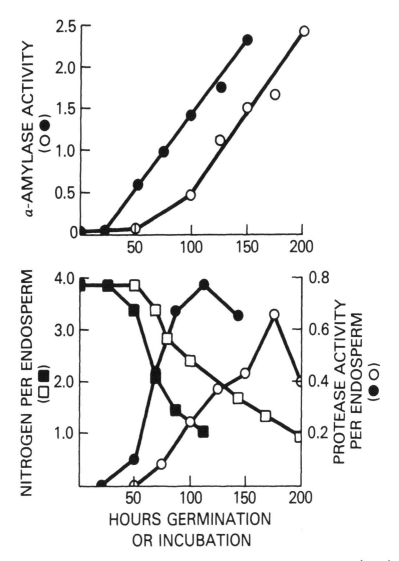

Fig. 1.   Development of α-amylase activity (●,o),
acid endopeptidase activity (●,o) and loss
of endosperm nitrogen (■,□) in normal (open
symbols) and excised endosperms (closed
symbols). After Harvey and Oaks (1974b, c).

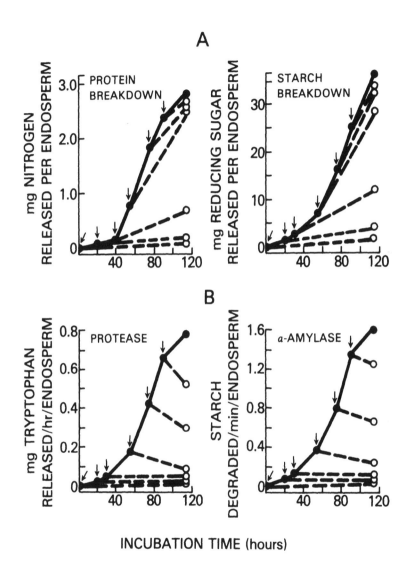

Fig. 2.   Effect of cycloheximide on the hydrolysis of
            endosperm storage products and on the devel-
            opment of α-amylase and acid endopeptidase
            activities.   Solid lines the control values
            (no cycloheximide); dotted lines are values
            obtained after treatment with cycloheximide
            (5 μg/ml).   After Harvey and Oaks (1974b).

can hydrolyze a wide range of protein substrates including urea-denatured zein and gliadin (Harvey and Oaks 1974a, b). Once the synthesis of the hydrolase activities (α-amylase or endopeptidase) is initiated there is no real control over the total production and, in fact, the enzymes are probably over-produced. For example, the synthesis can be inhibited at any time by the addition of cycloheximide (Fig. 2); however, by 30 hr. post imbibition the addition of cycloheximide has only a minor effect on the subsequent hydrolysis of endosperm protein or starch (Harvey and Oaks 1974b).

The increase in α-amylase, carboxypeptidase and endopeptidase activities with time in de-embryonated, H_2O-incubated maize endosperms, unlike the barley system which serve as a model for cereal work (Varner and Ho 1976), do not normally respond to gibberellic acid (Harvey and Oaks, 1974c, Winspear 1981). The results in Table I summarize the interaction of GA_3 and ABA in our system when we use fully ripe endosperm pieces and when we use endosperms from airdried seeds harvested 20 days past pollination. In the mature seed GA_3 has no effect on the induction α-amylase and endopeptidase whereas there is a clear response to GA_3 when immature endosperms are used. As with barley (Chrispeels and Varner 1967) ABA inhibits the development of the hydrolases and the inhibition is overcome by the addition of GA_3. Protein synthesis is involved as indicated by the imbibitory effect of cycloheximide. The role of RNA synthesis in the induction is less clear cut, as shown by the lack of inhibition by cordycepin. Our interpretation is that the ABA concentration is high relative to the GA in the immature seed and thus, when GA_3 is added it overcomes a natural inhibition. This hormonal balance would be sufficient to suppress precocious germination on the hydrolysis of endosperm reserves while the grain is still on the cob. In the mature seed the ABA content is probably low leaving an endogenous GA concentration adequate to induce the hydrolase activities. We are currently examining both the levels of GA and ABA in endosperm tissue and the capacity of GA_3 to induce the hydrolase activities.

2. Hydrolysis of Endosperm Proteins: Since zein and glutein are insoluble in water and since the hydrolases we have measured up to this point in time are water soluble, it is possible that other as yet unidentified peptide hydrolases are involved in the initial hydrolysis of the storage proteins. To examine this possibility we have treated dried endosperm powder with commercial peptide hydrolases The results show that both carboxypeptidase and pronase are required for a maximum

TABLE I: THE EFFECT OF GIBBERELLIC ACID (GA₃) ON THE
DEVELOPMENT OF α-AMYLASE AND THE PEPTIDE
HYDROLASES ON MAIZE ENDOSPERM PIECES.

	α-amylase	peptide hydrolases		
		endo	carboxy	amino
	(mg substrate)	(µg product)		
Mature seed				
Day 0	0	0.9	3.3	10.7
Day 6	22.6(100)	6.9(100)	11.0(100)	11.9(100)
Values Relative to Control				
ABA (2 µM)	35	41	65	97
GA₃(30 µM)	146	106	96	84
ABA + GA₃	110	84	94	78
Cycloheximide (18 µM)	0	8	2	-
Condycepin (100 µM)	44	100	100	-

	α-amylase	endo	carboxy	amino
	(mg substrate)	(µg products)		
Immature Seed				
Day 0	0.5	3.7	4.7	47.6
Day 4	29.7(100)	1.0(100)	3.6(100)	5.6(100)
Values Relative to Control				
GA₃ (30 µM)	211	990	72	93

Units are mg substrate (α-amylase) or µg product (peptide hydrolases) per g.fresh weight in 1 min. Endosperm from mature seeds were excised and incubated for 6 days; from seeds harvested 20 days post pollination and then air dried (immature seed) for 4 days.

TABLE II: EFFECT OF COMMERCIAL PROTEASES ON THE RELEASE
OF NITROGEN FROM MAIZE ENDOSPERM POWDER.

	Powder + Pronase	Powder + Carboxy- peptidase	Powder + Pronase + Carboxy- peptidase
Zero Time:			
Total N	67.2 (46.6[*])	46.4	82.5
48 h.			
Total N:	67.2 (44.7[*])	46.4	82.5
α-NH$_2$N[**]	31.4	4.2	44.4
Proline:			
Powder[***]	N.D.	N.D.	N.D.
Endosperm[***] leachate	N.D.	-	-

[*] The total N in the initial powder was 46.6; after treatment with SDS most of the nitrogen was in the medium after 48 h. at 40°C. Units are in μmoles per g of powder; pH of the assay was 8.3.

[**] Values for powder + pronase, powder + carboxypeptidase, or powder + pronase + carboxypeptidase were corrected for values obtained by the commercial enzymes alone. (i.e. Total for pronase was 31.5, for carboxypeptidase was 5.3 and for both enzymes were 44.1)

[***] Recovery of proline from acid hydrolyzed endosperm powder or in hydrolyzed endosperm leachate (see Oaks and Beevers 1964) was 4.9 or 5.2 respectively. N.D. means not detected in amino acid analysis.

digestion. Proline is not present in the enzyme hydrolyzed digests even though it represents 11% of the amino
acid content of the endosperm. After a 48 h. digestion
the level of $\alpha\text{-NH}_2\text{N}$ in the medium is almost equal to the
total nitrogen, results which suggests a complete hydro-
lysis of the endosperm protein. If we hydrolyze the
endosperm powder, or the enzyme hydrolyzed extracts,
with 6N HCl at 120°C for 4 h. proline is detectable Thus
the action of the commercial peptide hydrolases is
inhibited by the presence of proline in the primary
structure. A similar restriction of endogenous peptide
hydrolase action is also suggested by the absence of
free proline in unhydrolyzed leachates obtained from
endosperm pieces (Oaks and Beevers 1964).

REFERENCES

Blohm, D. 1972. Untersuchungen zur Aminosaure-biosyn-
these und Stickstoff-Assimilation in Keimpflanzen-
wurzeln Dissertation, Math. Nat Fakultat, Humboldt-
Universitat, Berlin.
Chen, S.S.C. and L.L. Chang 1972. Does gibberellic acid
stimulate seed germination via amylase synthesis?
Plant Physiol. 49, 441-2.
Chrispeels, M.J. and J.E. Varner 1967. Hormonal control
of enzyme synthesis: On the mode of action of
gibberellic acid and abscisin in aleurone layers of
barley. Plant Physiol. 42, 1008-1016.
Folkes, B.F. and E.W. Yemm 1958. The respiration of
barley plants. Respiration and metabolism of amino
acids and proteins in germinating grain. New
Phytol. 57, 106-131.
Goldbach, H. and G. Michael 1976. Abscisic acid content
of barley grains during ripening as affected by
temperature and variety. Crop Sc. 16, 797-9.
Feller, U., and T.-S.T. Soong and R.H. Hageman 1978.
Patterns of proleolytic enzyme activities in different tissues of corn. (Zea mays L) Planta 140,
155-162.
Harvey, B.M.R. and A. Oaks (1974a). Characteristics of
an acid protease from maize endosperm. Plant
Physiol. 53, 449-452.
Harvey, B.M.R. and A. Oaks (1974b). The hydrolysis of
endosperm proteins in Zea mays. Plant Physiol. 53,
453-457.
Harvey, B.M.R. and A. Oaks (1974c). The role of gibberellic acid in the hydrolysis of endosperm reserves
in Zea mays. Planta 121, 67-74.
McDaniel, S., J.D. Smith, H.J. Price 1977. Response of
viviparous mutants to abscisic acid in embryo culture. Maize Genetics Cooperation News Letter 51,
85-86.
Mikola, J. 1982. Proteinases, peptidases, and inhibitors of endogenous preoteases in germinating seeds.

Proc. Phytochem. Soc. (Europe) in press.

Oaks, A. 1965. The regulation of nitrogen loss from maize endosperm. Can. J. Bot. 43, 1077-1082.

Oaks, A. and H. Beevers 1964. The requirement for organic nitrogen in Zea mays embryos. Plant Physiol. 39, 37-43.

Sawhney, R. and J.M. Naylor 1979. Dormancy studies in seed of Avena fatua. 9. Demonstration of genetic variability affecting the response to temperature during seed development. Can. J. Bot. 57, 89-63.

Sodek, L. and C.M. Wilson 1973. Metabolism of lysine and leucine derived from storage protein during the germination of maize. Biochem. Biophys. Acta. 304, 353-362.

Srivastava, H.S., A. Oaks and I.L. Bakyta 1976. The effect of nitrate on early seedling growth in Zea mays. Can. J. Bot. 54, 923-929.

Toole, E.H. 1924. The transformations and course of development of germinating maize. Am. J. Bot. 11, 325-352.

Varner, J.E. and D.T. Ho 1976. The role of hormones in the integration of seedling growth. In The molecular biology of hormone action. (Ed. J. Papaconstinou (Ed.) 173-194.

Wall, J.S. and J.W. Paulis 1978. Corn and sorghum grain protein. In Adv. in Cereal Science and Technology (Ed. Y. Pomeranz) 2, 135-219.

Walton, D.E. 1980/81. Does ABA play a role in seed germination? Israel Journal of Biology 29, 168-180.

Winkler, U. and W.J. Schon 1979. Enzymatic hydrolysis of seed protein: A procedure for evaluating the nitritive value. In: Seed protein improvement in cereals and grain legumes. International Atomic Energy Agency. Vol. I, 343-351.

Winspear, M.J. 1981. Peptide hydrolases in maize endosperm. M.Sc. Thesis, McMaster University, 1-113.

Fusarium moniliforme as the Cause of Pre-Harvest Sprouting of Maize in Mexico

J. Galindo and S. Romero, *Centro de Fitopatología, Chapingo, México*

INTRODUCTION

Since the early 70's, preharvest sprouting or premature germination of maize has become a serious disease in some regions of the State of Puebla and Tlaxcala in Mexico. The economic losses are significant for the farmer because maize ears with some germinated seeds are not accepted as food by pigs, fowls, and other animals.

Up to 73 % of preharvest sprouted ears has been reported in some corn fields of the State of Puebla, and in such ears from 5-23 % germinated seeds has been found (6). The incidence of the disease increased when maize plants were fertilized with ammonium sulfate (3).

The occurrence of preharvest sprouting of maize has been recorded in the U.S.A., France, Venezuela and México (2,8,4,3). Most of the research done on this disorder has had the aim of determining its etiology. In U.S.A., genetic factors were usually considered the causal agent (2,5,7). It has also been shown that the fungus Diplodia maydis is able to induce preharvest sprouting (1). In France, evidence was obtained indicating that manganese deficiency stimulated preharvest sprouting (8) In Mexico negative results were obtained with the fungi Fusarium moniliforme and F. roseum isolated from infected ears of maize (3).

Since premature germination appears to be spreading to some other regions of Mexico it was considered desirable to re-investigate the hypothesis that F. moniliforme could be the causal agent, in spite of the negative evidence previously mentioned. The bases for this decision were: i) The close association which exists between germinated kernels and the presence of F. moniliforme;ii) The fact that Fusarium species produce harmful substances

to animals and the repulsion the animals feel for germinated kernels; and iii) The knowledge that in many of the syn
dromes caused by plant pathogenic Fusaria there is eviden
ce that an unbalance of growth regulating substances takes place during pathogenesis. Therefore, it was suspected that F.moniliforme in intimate association with corn
tissues could induce premature germination by the produc
tion of gebberellins.

MATERIALS AND METHODS

Maize ears with sprouted kernels were collected from
Puebla fields. The sprouted kernels associated with Fusarium infections were detached, surface desinfested with
sodium hypochloride and placed on Petri dishes with pota-
to-dextrose-agar medium (PDA). Most of the fungus colo-
nies isolated from internal tissues of sprouted kernels
corresponded to the species F. moniliforme,and they were
used to inoculate stigmata of maize plants under field
conditions, or ears under laboratory conditions.

The inoculum for the field inoculations was prepared
by washing off the spores formed by 8 F.moniliforme colo
nies of 9 cm diameter into 1000 ml water. The spore sus
pension was sprayed on the stigmata of 20 "criollo"plants
in Tlamanilco, Puebla. Ninety days after inoculation,
the preharvest phenomenon was recorded and also the germinated seeds of each ear. Since 2 plants were lost,the
data were taken only from the remaining 18 inoculated
ears. As control, 18 non-inoculated ears were taken at
random and examined in the same manner.

Under laboratory conditions 3 separate experiments
were performed, using only one isolate of F. moniliforme.
For the first experiment 20 "criollo" healthy ears were
selected from a Tlamanilco field in Puebla. Then, with a
spore suspension, 10 ears were sprayed, placed in two plas
tic trays, and each tray covered with a plastic bag. The
other 10 ears were only sprayed with sterile distilled
water, placed in two trays and covered with plastic bags.
The incubation temperature was about 22°C. For the second and third experiments only two ears obtained from a
local market were used for each treatment. The germination data were taken 5 to 8 days after the first kernels
started to germinate, usually, 12 to 15 days after the
inoculation.

RESULTS

From the 18 inoculated ears under field conditions,
12 presented premature germination, whereas only 3 ears
of the control did so. The total number of germinated
seeds in the inoculated ears was 29, and 7 seeds in the
control (Table 1).

TABLE 1.
Preharvest sprouting in maize ears inoculated with Fusarium moniliforme and in non inoculated ears, under field
conditions.

Treatment	Ears used	Ears with germination		Total germinated seeds
		amount	percentage	
Stigmata inoculated	18	12	66	29
Control	18	3	16	7

In the first experiment under laboratory conditions,
a total number of 1664 seeds germinated in the 10 inoculated ears, whereas only 82 seeds in the control ears.
In the second experiment, 445 seeds germinated in the two
inoculated ears and only 38 in the control ears. In the
third experiment 331 seeds germinated in inoculated ears
whereas only 12 in the non inoculated ears. In all experiments a total of 2440 germinated seeds were obtained
in the inoculated ears, whereas only 132 germinated seeds
in the non-inoculated ears, (Table 2).

TABLE 2.
Seed germination induced by inoculating maize ears with
Fusarium moniliforme spores under laboratory conditions.

Experiment	Treatments	Ears	Germinated Seeds
I	Inoculated	10	1664
	Control	10	82
II	Inoculated	2	445
	Control	2	38
III	Inoculated	2	331
	Control	2	12

DISCUSSION

Apparently the environmental conditions were not favorable for the preharvest sprouting in the field experiment, as shown by the very low number of germinated kernels. However, the 50 % difference observed between the
inoculated and the non-inoculated ears (66 vs. 16 %) is
indicative that Fusarium spores applied to the stigmata
induced premature germination.

On the other hand, in the laboratory experiments whe
re the environmental conditions were controlled, the degree of sprouting was very high and the large difference between the inoculated and non-inoculated ears was ve
ry consistent in all three experiments. Therefore, it is

216

evident that the Fusarium infection was responsible for the high germination occurring on the inoculated ears. This evidence strongly suggests that Fusarium moniliforme is the causal agent of the preharvest sprouting of maize in Mexico.

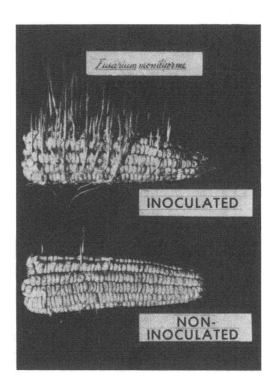

Fig. 1. Seed germination induced by Fusarium moni-liforme on maize ears.

REFERENCES

1. Calvert et al 1969. Vivipary in Zea mays induced by Diplodia maydis. Phytopathology 59: 239-240.

2. Eyster, W.H. 1931. Vivipary in maize. Genetics 16: 574-590.

3. Félix Gastélum Rubén y Sebastián Romero C. Etiología de la germinación prematura del maíz en Huaman tla, Tlaxcala. Agrociencia 43:81-87.

4. González, N.C. 1976. Germinación de granos inmadu-
    ros en mazorcas de maíz. Agronomía Tropical
    26 (A): 359-362.

5. Mangelsdorf, P.C. 1930. The inheritance of dormance
    and premature germination in maize. Genetics
    15:462-494.

6. Ortíz C. Joaquín. Personal communication

7. Robertson, D.S. 1955. The genetics of vivipary in
    maize. Genetics 40:745-760.

8. Routchenko, W. et Soyer, J.P. 1972. Cause de la ger
    mination sur Plante de grains inmatures de maiz.
    Donnés complementaires. Ann. Agron. 23 (4):
    445-459.

ACKNOWLEDGEMENT

We thank Sra. Teresa Aguilar for technical assistance.

Section III

Plant Breeding and Genetic Aspects of Pre-Harvest Sprouting

Sprouting Variability in Diverse *Triticum* spp. Germplasms

I. L. Gordon, *Massey University, Palmerston North, New Zealand*

SUMMARY

Ninety-seven diverse lines from several Triticum spp were tested in a warm ripening environment for alpha amylase, grain appearance and bulk density after controlled wetting. Grain colour and period of ripening also were measured. Descriptive heritabilities over all species were 0.44, 0.45, 0.73, 0.83, 0.90 respectively. Differences amongst lines generally formed gradations from one extreme to the next, without marked separation points (except for grain colour). The white wheats Spoetnik and Stewart had low levels of alpha-amylase and deterioration of appearance, and may be regarded as semi-resistant to sprouting.

INTRODUCTION

Much of our knowledge on germinability in wheat grain has been obtained from restricted genetic bases of elite lines and cultivars of a few countries. For example, after examining the papers of Nilssohn-Ehle(1914), Gfeller and Svejda(1960), Wellington and Durham(1958), Derera et al.(1976), Freed et al.(1976), LaCroix et al.(1976), and Gordon(1980(a)), forty-eight common-wheat lines had been studied, of which at least four appeared to be related. Five of these lines had been used in more than one study. Of the total lines used, thirty-eight were named cultivars.

Another restriction in much of this work has been that grain germinative maturity may have been confounded with dormancy, so that lack of germination may not have been an unequivocal indication of resistance to sprouting damage. Gordon et al.(1979) and Gordon(1979) showed that base-germinability ("maturity"), empirical dormancy and harvest ripeness were separable aspects of grain ripening. Belderok(1968) drew attention to the effects of temperature on germinability; and Gordon(1980(c)) reported work (with R.J.Cross) which showed that "maturity" was delayed greatly (relative to harvest ripeness) by cool ripening, whereas dormancy was affected less.

The present paper reports variability for sprouting damage traits in wheat germplasm from several Triticum spp. Furthermore, most of the lines are not studied commonly for sprouting damage, thereby widening the gene-base about which some sprouting information is known. Also, the results have been obtained under control for harvest ripeness, in order to minimise the confounding of dormancy with immaturity.

MATERIALS AND METHODS

Ninety-seven lines from the University of Sydney wheat collection were grown in three randomised blocks, in a warm ripening field environment in the northern New South Wales hard-prime wheat belt. The plots were single rows of 10 m length, spaced 0.5 m apart. These lines came from common wheat (Tr.aestivum aestivum), durum (Tr.turgidum durum) and other Triticum species. Table 1 indicates the numbers of white-, intermediate- and red-grained lines in each species group; together with an indication of whether they had long, medium or short periods of ripening of the grain. Two lines were duplicated (Mendos and Thatcher), the duplicates being sheltered from rain during grain ripening in order to adjust for uncontrolled sprouting damage. Mendos represented a white-grain, sprouting-susceptible control, while Thatcher was a red-grained sprouting-resistant control.

Table 1. Numbers of lines tested for sprouting in 5 spp groups, showing grain colour and ripening period

Grain colour:	white				intermediate				red				Tot
Ripening: Species group:	L	M	S	T	L	M	S	T	L	M	S	T	
Tr. aest. aest.	14	11	2	27	0	0	1	1	9	8	7	24	52
Tr. aest. other	2	1	0	3	0	1	0	1	1	1	1	3	7
Tr. turg. durum	1	4	0	11	1	3	1	5	0	2	3	5	21
Tr. turg. other	0	1	0	1	0	0	1	1	1	4	3	8	10
Tr. other	0	1	0	1	0	0	0	0	1	3	2	6	7

L=long, M=medium, S=short. T=Total.

Grain samples from each plot were tested for moisture concentration in order to determine harvest ripeness (Gordon et al., 1979(a)), which was defined at 12.5 % moisture, fresh-weight basis. Hand harvesting of primary heads took place at about 3 weeks after harvest ripeness. An aliquot of heads was kept dry. The remainder were sprayed for 5 minutes every half-hour for 48 hours while standing upright in clamps in a glasshouse. The spray details were similar to those described previously for a rain simulator (Gordon et al., 1977). Heads were dried in the stands for 24 hours, the grain was threshed by hand, and the grain was re-dried at 60°C for 8 hours. The dry aliquot of heads was also hand-threshed.

The characters measured were as follows:
    1 graincoat colour (score from 1 to 10, details follow);
    2 days from ear-peeping to wetting;
    3 alpha-amylase wetted grains (log EU) Gordon et al. 1977;
    4 alpha-amylase unwetted grains (log EU);
    5 grain appearance wetted (score from 1 to 11, Gordon et al. 1977;
    6 grain appearance unwetted (score 1 to 11);
    7 grain bulk density wetted;
    8 grain bulk density unwetted.

Graincoat colour was judged against a set of grains from ten lines which represented a graded colour series from pale "white" (score 1) to deep brown-red (score 10). All colour was judged after 1 hour immersion in 5% (w/v) aqueous sodium hydroxide (Quartley and Wellington, 1962). The ten lines had been chosen from the modal ranks of a set of thirty lines judged independently by four people. Insofar as the colours appeared to change from one standard to the next by similar amounts, the score was substantially an interval scale. The ten standards, and the scores they represented, were: 1 W676 "Russian" C4458, 2 W1538 Stewart, 3 W2572 world collection 54/503, 4 Gamenya, 5 W273 "Indian dwarf", 6 W2607 IRN59-88, 7 W2979 Nainari 60, 8 W2530 Benzengubskaia K40583, 9 W2906 and 10 W2888. (The "W" numbers were from the University of Sydney wheat collection.)

Data were analysed using standard univariate analyses of variance for an appropriate model. The sources of variability were examined through variance components, and the relative sizes of the contributions from meso-environment (blocks) and genotypes were estimated as ratios to the micro-environment (error). Descriptive heritabilities were estimated, as

$$\sigma_G^2 / (\sigma_B^2 + \sigma_G^2 + \sigma^2).$$

Their standard errors were obtained using procedures similar to those given in Gordon et al. (1972). Differences amongst means were tested using Duncan's Multiple Ranges, at the 5% level of significance. The general effect of sprouting for each relevant character was tested for significance with t-tests between the pair of means for wetted and unwetted in each case. A test for field sprouting (from rain) was made with a t-test

224

Table 2. Grand means, their standard errors and their
significance groups across species (P=0.05) for
each character, coefficient of variation (CV),
and t-test of general sprouting effect, for each
species group and overall.

Character:	1	2	3	4	5	6	7	8
Statistic:								

Tr. aest. aest.

Mean	4.7	73.4	2.27	1.61	6.4	3.3	0.634	0.676
se. mean	0.08	0.23	0.031	0.055	0.11	0.09	.0024	.0046
Sig set	b	a	b	b	a	a	b	a
CV(%)	20.9	3.9	17.2	43.7	21.6	36.4	4.9	8.7
Sprtg t	-	-	10.48	**	21.81	**	8.09	**

Tr. aest. other

Mean	4.5	73.4	1.95	0.77	4.9	2.5	0.613	0.653
se. mean	0.17	0.20	0.085	0.332	0.27	0.18	.0127	.0102
Sig set	b	a	c	c	b	bc	bc	b
CV(%)	17.8	1.3	20.0	197.9	25.7	33.0	9.5	7.2
Sprtg t	-	-	3.44	**	7.40	**	2.46	**

Tr. turg. durum

Mean	4.5	66.0	2.43	1.90	6.7	3.3	0.602	0.659
se. mean	0.10	0.65	0.046	0.052	0.20	0.18	.0047	.0045
Sig set	b	b	a	a	a	a	c	b
CV(%)	17.6	7.8	15.0	21.5	23.9	43.5	6.2	5.5
Sprtg t	-	-	7.63	**	12.64	**	8.76	**

Tr. turg. other

Mean	6.4	66.0	2.41	1.57	6.4	3.0	0.592	0.649
se. mean	0.26	0.26	0.114	0.060	0.37	0.19	.0120	.0084
Sig set	a	b	ab	b	a	ab	c	c
CV(%)	21.8	2.1	25.9	20.9	31.5	35.7	11.1	7.1
Sprtg t	-	-	6.52	**	8.17	**	3.89	**

Tr. other

Mean	6.9	67.0	1.20	1.06	4.0	2.3	0.655	0.676
se mean	0.22	0.81	0.372	0.102	0.31	0.21	.0083	.0052
Sig set	a	b	c	c	b	c	a	a
CV(%)	14.5	5.5	141.7	43.9	36.0	41.8	5.8	3.5
Sprtg t	-	-	0.36	NS	4.54	**	2.14	*

Over all spp

Mean	5.0	70.6	2.22	1.57	6.2	3.1	0.623	0.668
se mean	0.06	0.20	0.036	0.041	0.09	0.07	.0023	.0029
CV(%)	20.0	4.8	27.5	44.5	24.5	38.3	6.3	7.6
Sprtg t	-	-	11.91	**	27.19	**	12.16	**

NS=not sig,(NS)=sig at 10%,*=sig at 5%,**=sig at 1%.
Notes - a: means with same letter are NS from one another
b: see Methods for character identifications.

---

between covered and uncovered controls.

RESULTS

The central values obtained for each character (the grand means) are presented in Table 2 for each species group, and overall. Coefficients of variation are given also. Significant differences across species groups are given in Table 2 as well, together with the t-tests between wetted and unwetted means for alpha-amylase, grain appearance and bulk density.
The effects of field rain on the characters were mostly non-significant, as shown in Table 3.

Table 3. T-tests between open and rain-sheltered plots of Mendos and Thatcher controls.

Character:	1	2	3	4	5	6	7	8
Control:								

Mendos	0.41	6.05	0.25	0.41	1.62	0.68	0.47	0.10
	NS	**	NS	NS	NS	NS	NS	NS
Thatcher	0.41	1.09	0.10	0.50	0.54	1.02	0.07	4.79
	NS	NS	NS	NS	NS	NS	NS	**

The sources of variation in the germplasm have been indicated in Table 4, where the meso-environment (blocks) and genotype variances have been expressed as ratios to the micro-environment variance (error). The significance of each mean-square is shown also. This analysis of the variance is concluded with estimates of descriptive heritability.
The significant differences amongst the line means revealed a large amount of information about line sprouting potential, but it is too extensive to provide here. Some of the principal findings for each character are given in Table 5. These include the total number of significance groups amongst the means, the number of such groups in the two extremes of the ranked means, the percentage of red-grained wheats in the two extremes, and the percentage of long-ripening wheats in these extremes. For the two larger groups, each of the "extremes" consisted of one third of the lines. The "extremes" for the smaller groups were one half each.
Some white-grained wheats showed some degree of sprouting resistance. The best white wheats are given in Table 6 for common-wheat, durum and overall. The Table also shows the worst red-grained wheats, and the minima and maxima for the three characters.

Table 4. Relative sizes of variances due to blocks and
genotypes compared with error,their significances
and descriptive heritabilities

Character: Statistic:	1	2	3	4	5	6	7	8
Tr. aest. aest.								
Blocks	0.02	0.03	-0.01	-0.02	0.03	0.08	0.02	0.01
ratio	NS	(NS)	NS	NS	(NS)	**	NS	NS
Genotype	5.60	11.51	1.82	0.55	1.02	0.47	3.68	0.72
ratio	**	**	**	**	**	**	**	**
Herit.	0.85	0.92	0.65	0.36	0.50	0.30	0.78	0.42
se. hrt.	.032	.018	.063	.087	.077	.084	.043	.083
Tr. aest. other								
Blocks	-0.10	0.17	0.05	-0.02	0.17	-0.07	-0.10	0.06
ratio	NS	NS	NS	NS	NS	NS	NS	NS
Genotype	8.70	140.3	1.88	0.14	0.50	0.05	3.16	5.11
ratio	**	**	**	NS	(NS)	NS	**	**
Herit.	0.91	0.99	0.64	0.13	0.30	0.05	0.78	0.83
se. hrt.	.055	.005	.163	.230	.211	.232	.118	.091
Tr. turg. durum								
Blocks	0.35	-0.03	0.00	-0.03	-0.04	-0.03	-0.03	-0.02
ratio	**	NS	NS	NS	NS	NS	NS	NS
Genotype	5.61	2.62	1.76	1.34	0.32	0.41	2.70	2.59
ratio	**	**	**	**	*	*	**	**
Herit.	0.81	0.73	0.64	0.58	0.25	0.29	0.74	0.73
se hrt.	.065	.082	.101	.113	.142	.140	.081	.083
Tr. turg. other								
Blocks	-0.05	-0.02	-0.09	0.03	0.05	-0.08	-0.08	-0.06
ratio	NS	NS	NS	NS	NS	NS	NS	NS
Genotype	0.88	45.24	-0.05	1.02	0.58	0.27	0.18	0.33
ratio	**	**	NS	**	*	NS	NS	NS
Herit.	0.48	0.98	-0.05	0.50	0.36	0.23	0.16	0.26
se. hrt.	.181	.011	.186	.172	.187	.207	.207	.204
Tr. other								
Blocks	0.25	0.07	0.05	-0.02	0.52	0.55	-0.13	0.00
ratio	NS	NS	NS	NS	*	*	NS	NS
Genotype	3.66	6.11	0.01	0.41	-0.12	0.02	3.31	9.27
ratio	**	**	NS	NS	NS	NS	**	**
Herit.	0.75	0.85	0.01	0.30	-0.09	0.01	0.79	0.90
se. hrt.	.125	.080	.201	.232	.118	.138	.113	.055
Over all spp								
Blocks	0.04	0.00	0.00	0.00	0.03	0.06	0.01	0.01
ratio	**	NS	NS	NS	*	**	NS	(NS)
Genotype	4.94	8.68	0.79	0.64	0.85	0.45	2.66	1.25
ratio	**	**	**	**	**	**	**	**
Herit.	0.83	0.90	0.44	0.39	0.45	0.30	0.73	0.55
se. hrt.	.027	.017	.061	.063	.060	.063	.039	.054

```
------------------------------------------------------------
Table 5. Principal features of means analyses at P=0.05
------------------------------------------------------------
Character:                  1   2   3   4   5   6   7   8
Statistic:
------------------------------------------------------------
Tr. aest. aest.
No. sig. sets               3  19  14   8   9   6  21   9
No. sets in top extreme     1   8   7   8   4   6  10   7
No. sets in bottom          1   9  11   8   7   4  12   9
% red grain, top extreme  100  28  28  17  22  33  67  72
% red grain in bottom       0  72  78  83  89  67  33  39
% long ripeners in top     22 100  67  67  78  56  22  50
% long ripeners,bottom     61   0  28  33  22  44  44  33

Tr. aest. other
No. sig. sets               3   4   3   1   1   1   3   4
No. sets in top extreme     1   2   2   1   1   1   2   2
No. sets in bottom          2   2   1   1   1   1   3   2
% red grain, top extreme  100  33   0   0   0   0  67  67
% red grain in bottom       0  67 100  67 100  67  33   0
% long ripeners in top     33 100  33  33  33  67   0  67
% long ripeners,bottom     67   0  33  67  33   0  67  67

Tr. turg. durum
No. sig. sets               6   8   5   7   4   5   9   8
No. sets in top extreme     3   3   3   4   3   4   5   3
No. sets in bottom          3   5   5   5   4   5   4   5
% red grain,top extreme    71  14  14  14  29  43  43  29
% red grain in bottom       0  29  29  29  14  14   0  14
% long ripeners in top     14  29  14  14  14  14  14  14
% long ripeners,bottom     14   0   0   0   0   0   0   0

Tr. turg. other
No. sig. sets               2   7   1   3   3   1   1   1
No. sets in top extreme     1   5   1   2   2   1   1   1
No. sets in bottom          2   3   1   2   3   1   1   1
% red grain,top extreme   100 100 100  60  80  80  60  80
% red grain in bottom      60  60  60 100  80  80 100  80
% long ripeners in top     20  20   -   -   -  20   -   -
% long ripeners,bottom      -   -  20  20  20   -  20  20

Tr. other
No. sig.sets                3   3   1   1   1   1   3   5
No. sets in top extreme     2   2   1   1   1   1   2   3
No. sets in bottom          2   2   1   1   1   1   1   2
% red grain,top extreme   100  67  67  67  67  67  67 100
% red grain in bottom      67 100 100 100 100 100 100 100
% long ripeners in top     33  33   0  33   0   -   0  33
% long ripeners,bottom      -   -   0   -   0  33   0   -
------------------------------------------------------------
Note: "-" = only one long-ripener in the set.
```

DISCUSSION

Alpha-amylase increased significantly with sprouting in all species groups except the oddments, which consisted of various crosses and Tr.monococcum. The two sets of Tr.turgidum produced the most alpha-amylase. These two groups also suffered badly from visible deterioration, as did common-wheat. But, all species groups deteriorated to some extent in grain appearance. Bulk density decreased significantly with sprouting too, the trends across species being similar to those of alpha-amylase.

None of the sprouting characters had any significant bias from field rain. The sheltered Mendos ripened more slowly than the open plot, and the sheltered Thatcher had a higher bulk density. But none of the other characters were affected significantly.

Table 6. Most sprout-resistant white wheats, most sprout susceptible red wheats, and character extremes, for two commercial species groups and overall.

Grain colour:	best white		worst red			
Extremes:					min	max
	Name	Mean	Name	Mean		
Character:						
Tr. aest. aest.						
3	Spoetnik	1.90	Mayo 64	3.00	1.08	3.14
5	Spoetnik	4.7	HD(M)1827	8.0	3.3	9.3
7	Brdd Kruger	.740	Tammi	.486	.460	.740
Tr. turg. durum						
3	Stewart	1.60	WC 54/383	2.67	0.69	3.05
5	Stewart	4.7	WC 54/383	8.0	3.7	9.0
7	Doubbi	.691	Gwalior	.579	.494	.697
Overall						
3	Stewart	1.60	W1042	3.03	1.60	3.14
5	Stewart	4.7	W1042	8.0	3.7	9.0
7	Brdd Kruger	.740	Tammi	.486	.391	.740

Variability from meso-environment (blocks) was consistently trivial. Genotypic variability was usually substantial, especially in the two larger samples of lines. It was often much larger than the residual variance. Heritabilities were mostly consistent from one species group to the next; but, again, the smaller samples of lines showed the greatest anomalies. These heritabilities were probably narrow-sense, because of the long-term selfing giving rise to most lines. The heritabilities may be biased upwards because of genotype-environment interactions which could not be partitioned out with only one macro-environment. It was

interesting to note that colour was not 100% heritable; and that the ripening-period had the highest heritability in the overall analysis.
Significance grouping of lines led to clear-cut separation into distinct types only in the case of grain colour. Mostly, intermediate grain-colour was grouped with red grain. For the other characters, series of overlapping significance groups occurred.
The pattern of associations between sprouting reactions and grain colour, or ripening period, was not easily discernable. Red wheats were well distributed over the whole range of values for all three sprouting characters, but there did appear to be a preponderance of them towards the less susceptible end. It was even more difficult to detect a trend for length of ripening period. A factor analysis may be helpful in assessing the patterns in these associations, and this is presented elsewhere.

REFERENCES

Belderok B. 1968. Seed dormancy problems in cereals. Field Crop Abstr. 21(3), 203-211.

Derera, N.F., McMaster, G.J., and Balaam L.N. 1976. Pre-harvest sprouting resistance and associated components in twelve wheat cultivars. Cereal Res. Commun. 4:173-179.

Freed, R.D., Everson, E.H., Ringlund, K., and Gullord, M. 1976. Seedcoat color in wheat and the relationship to seed dormancy at maturity. Cereal Res. Commun. 4:147-149.

Gfeller, F., and Svejda, F. 1960. Inheritance of post-harvest seed dormancy and kernel color in spring wheat lines. Can. J. Plant Sci. 40:1-6.

Gordon, I.L. 1979. Selection against sprouting damage in wheat. III. Dormancy, germinative alpha-amylase, grain redness and flavanols. Aust. J. Agric. Res. 30:387-402.

Gordon, I.L. 1980(a). Heritability of grain development traits associated with sprouting damage in wheat inbreds. Cereal Res. Commun. 8:185-192.

Gordon, I.L. 1980(b). Germinability, dormancy and grain development. Cereal Res. Commun. 8:115-129.

Gordon, I.L. 1980(c). Selection against sprouting damage in wheat: a synopsis. Proc. 5th Int. Wheat Genetics Symp., 954-962.

Gordon, I.L., Balaam, L.N., and Derera, N.F. 1979. II. Harvest ripeness, grain maturity and germinability. Aust. J. Agric. Res. 30:1-17.

Gordon, I.L., Byth, D.E., and Balaam, L.N. 1972. Variance of heritability ratios estimated from phenotypic variance components. Biometrics 28:4Ø1-415.

Gordon, I.L., Derera, N.F., and Balaam, L.N. 1977. Selection against sprouting damage in wheat. I. Germination of unthreshed grain with a standard wetting procedure. Aust. J. Agric. Res. 28:583-596.

LaCroix, L.J., Waikakul, P., and Young, G.M. 1976. Seasonal and varietal variation in dormancy of some spring wheats. Cereal Res. Commun. 4:139-146.

Nilssohne-Ehle, H. 1914. Zur Kentis der mit der Keinungsphysiologie des Weizens in Zusammenhang stehenden immeren Faktoren. Z. Pflanzenzuecht. 3:153-187.

Quartley, C.E., and Wellington, P.S. 1962. A test to distinguish between red and white grains of wheat treated with a seed dressing. J. Nat. Inst. Agric. Bot. 9:179-185.

Wellington, P.S., and Durham, V.M. 1958. Varietal differences in the tendency to sprout in the ear. Emp. J. Exp. Agric. 26:47-54.

Factor Analyses of Characters Useful in Screening Wheat for Sprouting Damage

I. L. Gordon, *Massey University,*
Palmerston North, New Zealand

SUMMARY

The factor-structure of principal-factors derived from three sprouting characters and six base characters was examined to identify the most prominent characters accounting for the dispersion in ninety-seven diverse wheat lines. Simple and partial correlations were compared, and led to divergent interpretations of associations amongst the characters. Multiple and canonical correlations between sprouting and base characters were estimated also. It was concluded that screening of lines for sprouting resistance should be based directly on sprouting attributes themselves, and that monitoring for harvest-ripeness was beneficial to obtaining valid results.

INTRODUCTION

The positive correlation between graincoat redness and apparent dormancy, which some wheat lines display, has been one association that has attracted attention in work on sprouting damage. Gordon (1979) discussed this association at length, and pointed out that the magnitude of the correlation estimate varied considerably amongst researchers, depending upon the sample of genotypes and environments used. Usually, the sample of both was fairly restricted, thereby limiting the generality of any one set of findings. This has led to some misleading statements asserting the ubiquity of dormancy in red-grained wheats.

Similar problems have attended another correlation often used in sprouting damage work, namely that between alpha-amylase and falling-number. Falling-number has been interpreted as a direct assay of alpha-amylase activity, but Gordon et al. (1977) concluded that falling-number was more accurately interpreted as a measure of endosperm degradation per se than as alpha-amylase activity. Several of their results led to this conclusion, and considered among them were marked differences between simple and partial correlations for the two characters. The point made earlier about the sample restrictions also applies for these characters, of course. It is clear, then, that care is needed when interpreting relationships amongst associated characters, especially when using the simple correlation to do so. A major problem with simple correlation is that all the associations amongst the characters studied affect the estimate obtained for any specific pair of characters in the set. Such statistics do not give the intrinsic association between two characters unencumbered by the other associations. The highest-order partial correlations do provide such information. Use of such statistics may alter considerably the interpretations made when examining sprouting processes via correlation estimates. This sort of examination is done quite often in biology, and amounts to an "intuitive" factor-analysis of a correlation matrix.

The present paper reports correlations amongst characters used in screening a diverse wheat germplasm for sprouting damage. Simple, partial, multiple and canonical correlations are given and discussed. Also, a principal-factor analysis follows, in order to identify those characters accounting for the bulk of the dispersion (variance and covariance) in the characters used.

MATERIALS AND METHODS

Ninety-seven lines from several _Triticum_ species were grown in a warm ripening field environment. Harvest ripe ears were subjected to controlled wetting for 48 hours, and were hand-threshed after drying. Uncontrolled sprouting was checked for. The characters measured were:

1 graincoat redness (score 1 to 10),
2 days from ear-peeping to wetting,
3 days from harvest-ripeness to wetting,
4 alpha-amylase, wetted grains (log EU),
5 alpha-amylase, unwetted grains (log EU),
6 grain appearance, wetted (score from 1 to 11),
7 grain appearance, unwetted (score from 1 to 11),
8 grain bulk density, wetted (g/ml),
9 grain bulk density, unwetted (g/ml).

Further details of the experiment and the characters are given in Gordon (1982).

Simple correlations were estimated amongst all character pairs, for each species group and overall (but only the latter is reported here). The highest-order partial correlations also were obtained for each character pair which involved a "sprouting" character (4,6,8) and a "base" character (1,2,3,5,7,9). In addition, for each sprouting character, the multiple correlation between it and all the base characters jointly was estimated. The final correlation obtained was the canonical correlation between the three sprouting characters jointly and all the base characters jointly. Only the first canonical variates were used, as they were always significant (at P=0.05), whereas successive canonical variates were significant for only some of the species groups. Lastly, unrotated principal factor analyses were done on the standardised characters. The partial correlations between the factors and the characters (the factor-structures) were obtained. These various methods and their purposes are presented in Steel and Torrie (1980) and Cooley and Lohnes (1971).

RESULTS AND ASSOCIATED DISCUSSION

The superficial associations (simple correlations) between the characters are given in Table 1 for all species groups overall. These correlations also are the dispersion (variances and covariances) of the standardised variables.

Table 1. Simple correlations (upper triangle) and partial correlations (lower triangle) over all species groups.

Character:	1	2	3	4	5	6	7	8	9
1	1	-.21 **	-.15 **	-.28 **	-.29 **	-.38 **	-.19 **	0.24 **	0.13 *
2	--	1	0.14 **	0.16 **	0.10 *	0.28 **	0.23 **	0.03 NS	0.08 (NS)
3	--	--	1	0.22 **	0.13 *	0.21 **	0.13 *	-.20 **	-.10 *
4	-.01 NS	0.04 NS	0.09 (NS)	1	0.51 **	0.53 **	0.37 **	-.39 **	-.14 **
5	--	--	--	0.32 **	1	0.43 **	0.41 **	-.24 **	-.09 (NS)
6	-.22 **	0.13 *	0.06 NS	0.24 **	0.06 NS	1	0.57 **	-.29 **	-.06 NS
7	--	--	--	0.05 NS	--	0.44 **	1	-.16 **	-.10 *
8	0.08 (NS)	0.09 (NS)	-.08 (NS)	-.26 **	-.03 NS	-.17 **	0.11 *	1	0.69 **
9	--	--	--	0.12 *	--	0.17 **	--	0.69 **	1

NS=not sig (NS)=sig at 10% *=sig at 5% **=sig at 1%
"--" = not estimated

Grain redness was only moderately associated with less alpha-amylase (wetted) and with better grain appearance (wetted), the latter being the stronger association.

Redness was weakly correlated with higher bulk density, both wetted and unwetted. Both ripening-period and period since harvest-ripeness had lesser correlations than redness with amylase (wetted) and appearance (wetted). These were also of opposite sign to those of redness. However, the relative trends in value amongst characters were similar to those for redness. The largest correlations were between each sprouting character and its own base level, but these were only medium in value. The correlation between base amylase and grain appearance (wetted) also was relatively large.

The intrinsic associations between the sprouting characters and the base characters were given by the highest-order partial correlations, also in Table 1. Grain redness was correlated at a low-moderate level with better grain appearance; but the correlation with wetted amylase or bulk density was trivial. All ripening-period and harvest-ripeness partial correlations were trivial. The latter was to be expected because of the monitoring of harvest-ripeness prior to harvest (Gordon et al.,1979). Mostly, the partials were lower in value than the corresponding simple correlations. The correlations between the sprouting characters and their respective base levels were the associations least changed.

The correlations between each sprouting character and all base characters jointly (Table 2) were higher in value generally than the pair-wise correlations.

```
---------------------------------------------
```
Table 2. Multiple correlations between each
"sprouting" character (4,6,8) and
all "base" characters; and the
first canonical correlation (C)
between all sprouting characters
and all base characters.
```
---------------------------------------------
```

Character:	4	6	8	C
Tr.aest.aest.	0.54 **	0.70 **	0.63 **	0.72 **
Tr.aest.other	0.81 *	0.68 (NS)	0.91 **	0.93 **
Tr.turg.durum	0.67 **	0.69 **	0.86 **	0.86 **
Tr.turg.other	0.47 NS	0.69 *	0.74 **	0.79 **
Tr.other	0.68 (NS)	0.84 *	0.90 **	0.92 **
Overall	0.57 **	0.67 **	0.73 **	0.73 **

```
---------------------------------------------
```

Table 3 Factor-structure of those principal-factors accounting cumulatively for at least 80% of the dispersion of the nine characters, with the highlights underlined.

% of disp.		Factor-structure correlations								
Character: Fr:		1	2	3	4	5	6	7	8	9
Tr.aest.aest.										
1	35.3	-.64	0.38	0.08	0.74	0.70	0.83	0.66	-.62	-.28
2	17.4	-.18	0.70	0.26	0.01	0.02	0.17	0.04	0.59	0.77
3	11.6	0.04	0.21	0.89	0.01	-.16	-.11	-.20	-.16	-.33
4	9.1	0.48	-.30	0.29	0.08	0.43	-.09	0.41	0.19	0.13
5	7.5	0.26	-.01	-.06	0.55	0.07	-.02	-.47	-.17	0.23
Tr.aest.other										
1	43.3	-.74	0.21	0.11	0.85	0.74	0.52	0.53	-.87	-.85
2	17.8	-.07	0.21	0.82	0.03	-.36	0.48	0.59	0.25	0.33
3	13.8	0.09	-.92	0.13	0.23	0.29	0.37	-.16	0.05	0.27
4	9.5	0.52	-.04	0.30	0.15	0.32	-.51	0.31	-.06	-.05
Tr.turg.durum										
1	30.7	-.27	0.41	0.41	0.72	0.72	0.59	0.54	-.63	-.52
2	22.2	0.56	0.35	-.09	0.14	0.16	0.48	0.61	0.67	0.68
3	13.9	0.25	0.64	0.75	-.29	-.21	-.27	-.06	0.00	-.09
4	10.5	-.09	-.16	0.33	0.43	0.43	-.42	-.29	0.18	0.38
5	8.6	0.70	-.37	0.04	0.02	0.14	-.18	0.13	-.20	-.21
Tr.turg.other										
1	25.4	-.18	0.48	0.24	0.69	0.28	0.85	0.77	0.09	0.28
2	19.4	-.30	-.13	-.04	-.33	0.22	-.06	-.09	0.88	0.84
3	15.3	0.81	0.16	-.20	0.15	-.67	-.06	0.19	0.30	0.24
4	12.1	0.15	0.45	0.80	0.09	0.08	-.35	-.26	-.01	0.14
5	11.3	0.21	0.64	-.47	-.19	0.50	-.18	0.02	0.07	-.14
Tr.other										
1	37.3	-.24	0.75	0.60	0.54	0.19	0.72	0.85	-.62	-.65
2	23.7	-.79	-.15	0.23	0.58	0.46	0.25	0.01	0.60	0.67
3	14.6	0.01	0.48	-.15	0.09	0.79	-.47	-.37	-.27	-.02
4	8.7	0.24	0.10	0.64	-.37	0.14	-.19	0.18	0.19	0.20
Over all spp										
1	34.3	-.55	0.32	0.38	0.75	0.67	0.78	0.65	-.59	-.37
2	17.2	-.06	0.47	-.03	0.05	0.14	0.28	0.29	0.70	0.81
3	11.0	0.25	-.53	-.64	0.17	0.37	0.07	0.25	0.04	0.07
4	9.4	0.47	-.36	0.64	0.18	0.13	-.02	0.04	0.07	0.17
5	8.7	0.58	0.36	-.09	-.12	-.23	0.09	0.42	-.08	-.23

Generally, alpha-amylase had the lower multiple correlations, but all were medium to moderately-high in value. For some of the smaller species groups, the low degrees of freedom led to some estimates being declared non-significant despite their apparently medium values. The three wetted characters together constituted a total sprouting damage syndrome, and so it was appropriate to find their joint association with the base characters. This was provided by the canonical correlations in Table 2. These showed that the entire set of sprouting characters was highly correlated with the whole set of base characters, three of which were putative "causal" attributes (1,2,3). The species groups all had high values of canonical correlation.

In Table 3, those principal factors which together accounted for at least 80% of the total dispersion in each species group (and overall) are given. The percentages of the dispersion accounted for by each successive factor are shown. This provides a measure of the "importance" of that factor. The factor-structure correlations (Table 3) indicated which characters were most highly associated with these factors, and thereby identified which were the principal attributes determining the variability and covariability in the system. Table 3 also shows the highlights in the factor-structures. A "highlight" was any character with a correlation greater than $|0.34|$, and in the top five of the list.

Grain redness was a highlight in the most important factor (the first) only for the two Tr.aestivum groups; and it appeared in the second or third factors for the other groups. Over all species, redness did not feature until the fourth factor (which accounted for only 9.4% of the dispersion). Ripening-period was prominent in the structure of the first factor in only two of the species groups, neither of which were major commercial ones. The period from harvest-ripeness never featured in a factor accounting for more than 14% of the dispersion. In all species groups (except the oddments), and overall, wetted alpha-amylase had high positive correlations with the first (most important) factor. Base alpha-amylase also was prominent in the first factors, except for the oddments and Tr.turgidum "other". Grain appearance had moderate-high correlations in most first factors, the exception being in Tr.aestivum "other". In that group, appearance featured in the second factor, but this accounted for only 0.41 of the dispersion accounted for by the first factor. Generally, drop in bulk density was moderately to strongly associated with the first or second (or both) factors. Several characters featured in more than one factor.

GENERAL DISCUSSION

In general, none of the simple correlations were strong, making it difficult to recognise patterns of association amongst the sprouting and base characters. Grain redness did not stand out as being strongly negatively correlated with any sprouting response. This was made even clearer when the partial correlations were considered. Neither ripening-period nor period since harvest-ripeness were strongly associated with any sprouting response, especially when the partials were examined. In view of the grain development descriptions of Gordon et al. (1979), this suggested that all lines ripened fully in this warm ripening environment; and that the monitoring for harvest-ripeness achieved its goal of minimising any confounding of immaturity with dormancy. The point made in the Introduction, that simple correlations may be misleading as estimates of the intrinsic associations amongst characters, was supported strongly by these results. The simple correlations involving sprouting characters and their respective base values were the least biased by other inter-correlations in the system, as shown by the relative similarity of them with the corresponding partials.

The multiple and canonical correlations provided additional interesting information concerning various aggregates of the characters. The square of the multiple correlation is the coefficient of determination in multiple regression. Therefore, the values obtained for the multiple correlation suggested that these base characters accounted for only a modest ammount of the variation in each sprouting character.

The factor-structures have indeed been a useful analytical tool. These results suggested that screening of lines for sprouting-resistance would be better based directly on sprouting attributes themselves, and each severally. This supports similar conclusions of Gordon et al. (1977, 1979) and Gordon (1979, 1980), which were based on physiological and heritability evidences. Grain redness and ripening-period were indicated as being less useful. As far as ripeness was concerned, it should be borne in mind that it was probably well-advanced (because of the warmth and the harvest-ripeness monitoring). In other circumstances, effects of immaturity may be more prominent.

The results reported here were based on a diverse germplasm of lines not commonly used for sprouting damage research. Therefore, the results may be quite representative of wheat generally, especially at the tetraploid and hexaploid levels, which were better represented in the sample.

238

REFERENCES

Cooley W.W. and Lohnes P.R. (1971). 'Multivariate data analysis', Wiley.

Gordon I.L.(1979). Selection against sprouting damage in wheat. III.Dormancy, germinative alpha-amylase, grain redness and flavanols. Aust.J.Agric.Res.30,387-402.

Gordon I.L.(1980). Heritability of grain development traits associated with sprouting damage in wheat inbreds. Cer.Res.Commun.8, 185-92.

Gordon I.L.(1982). Sprouting variability in diverse Triticum spp germplasms. Proc.3rd Int.Symp.Sprouting Damage in Cereals, in press.

Gordon I.L., Balaam L.N. and Derera N.F.(1979). Selection against sprouting damage in wheat. II.Harvest ripeness, grain maturity and germinability. Aust.J.Agric.Res.30,1-17.

Gordon I.L., Derera N.F. and Balaam L.N.(1977). Ditto. I.Germination of unthreshed grain, with a standard wetting procedure. Aust.J.Agric.Res.28,583-96.

Steel R.G.D. and Torrie J.H.(1980). 'Principles and procedures of statistics', 2nd edtn, McGraw-Hill.

Sprouting Resistance in Barley

B. L. Harvey, B. G. Rossnagel and
R. P. Muderewich, *University of Saskatchewan,
Saskatoon, Saskatchewan, Canada*

INTRODUCTION

Klages barley and several of its derivatives are very prone to
pre-harvest sprouting which is known to drastically reduce the quality of malting barley (Brookes, 1980). Normally, harvest conditions
in the traditional two row area of western Canada do not favor
sprouting even in susceptible varieties. However, in some years
and some locations wet harvest weather does occur. Thus the potential for serious problems does exist. Furthermore, the popularity
of Klages with the malting and brewing industry has led to the
cultivation of this variety in cooler, moister areas where the
probability of poor harvest conditions is much greater.

The basic difference between Klages and older malting varieties,
such as Betzes, is a much higher level of amylolytic enzyme activity.
This leads one to suspect that there is a relationship between
sprouting and amylolytic activity. This suspicion is supported by
the fact that Domen, the donor of high enzyme activity to Klages,
is also highly susceptible to sprouting. If this is correct it
leads to the question, can a suitable source of dormancy be found
which can be combined with high amylolytic activity?

MATERIALS AND METHODS

Wpg M143-2-3b a dormant line obtained from Dr. D.R. Metcalfe was
crosses to Norbert (susceptible), Fairfield (intermediate) and
S77433 (hulless). Spaced planted F_2's and F_3 head rows were grown
in the field in 1980 and 1981 respectively. Approximately 300
plants were harvested at random from the F_2 generation of each of
the three crosses. Individual plants were also harvested from each
of the four parents. In the F_3 generation 10 plants were harvested
individually from each row and the remainder of the row was bulk
harvested. Seed was threshed immediately and stored at $-19^{\circ}C$ until
germination tests were conducted.

A portion of the seed of the WM143-2-3b/Norbert cross was stored
at $+18^{\circ}C$ for eight months and then micromalted and analyzed for
α-amylase activity and diastatic power.

The analysis of the F_2 data of the S77433/WM143-2-3b cross revealed that germination of all hulless segregates was high whereas

the germination of hulled segregates varied from 0 to 95%. Therefore several experiments were conducted to investigate the role of the hull in sprouting resistance. Seed of WM143-2-3b was punctured through the lemma and palea using a dissecting needle and then germination tested. Seeds of WM143-2-3b were dehulled by hand and then germinated. Seeds of Norbert were germinated in the same petri dish as seeds of WM143-2-3b in the ratios of 1:4, 1:9 and 1:19. Hulls of WM143-2-3b, and S77431 (a sister line of S77433) were collected and used to produce aqueous extracts. These extracts were then tested to determine their effects on germination of the four parents.

RESULTS AND DISCUSSION

Germination in each of the F_2 populations varied from 0 to 100% and the germination of the parents in each case were WM143-2-3b 15%, Norbert 90%; WM143-2-3b 24%, Fairfield 49%; WM143-2-3b 18%, S77433 83% (Figs. 1,2 and 3). The growing season in 1981 was hot and dry thus the germination of F_3 segregates and parents was much higher than in 1980. For example the mean germination for Fairfield was 49% in 1980 and 94% in 1981, WM143-2-3b 20% in 1980 and 53% in 1981. This undoubtedly reduced the heritability.

Heritability was calculated from the regression of F_2 and F_3 data for the crosses WM143-2-3b/Norbert and Fairfield/WM143-2-3b. The heritabilities were 23.2% and 23.0% respectively. These values are higher than those reported by Mueller (1964) for F_2 and subsequent generations and lower than values reported for F_3 and subsequent generations.

The storage period of eight months was sufficient for the dormancy to disappear from the WM143-2-3b/Norbert cross thus germination was high in all of the F_2 plants which were micromalted. The correlations between germination (at harvest) and α-amylase and diastatic power of the malted grain were $r = 0.485**$ and $r = 0.32**$ respectively. This is in agreement with the findings of Gordon (1973). While these findings indicate that there is a negative relationship between sprouting resistance and amylolytic activity it should be possible to select genotypes which combine sprouting resistance with high amylolytic activity. We have selected F_5 lines from this cross which appear to combine these features. Dr. D.R. Metcalfe (personal communication) has also obtained such lines from a cross involving Maris Concorde and Klages derivatives from his breeding program.

As indicated above, all of the hulless segregates from the F_2 of the cross S77433/WM143-2-3b had high germination, whereas the hulled segregates varied from 0% to 95%. Thus the dormancy factor appears to reside in the hulls. Germination of the hand-dehulled seed of WM143-2-3b supports this observation. The germination of the dehulled seed was 100% as compared with 5% for hulled controls from the same lot. Several authors (Takahashi 1980, Corbineau 1980 and Dunwell 1981) have reported similar observations.

Seed of WM143-2-3b which was punctured with a dissecting needle had a lower germination than control seeds from the same lot (45.5%

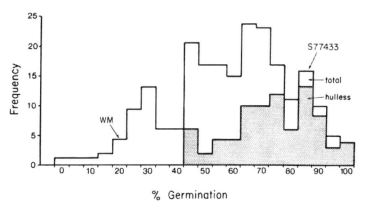

Figure 1. Germination of F_2 plants from the cross S77433/WM143-2-3b

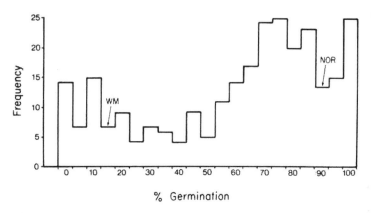

Figure 2. Germination of F_2 plants from the cross WM143-2-3b/Norbert

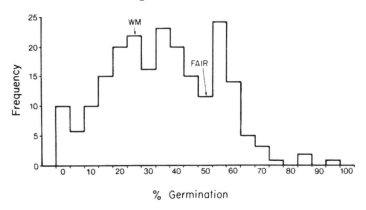

Figure 3. Germination of F_2 plants from the cross Fairfield/WM143-2-3b

and 58.3% respectively). Thus the effect of the hulls on dormancy is apparently not a physical effect. This does not agree with the observation of Dunwell (1981).

Table 1. The effects of aqueous hull extracts of S77431 on the germination of the four parents.

	Control	*5 g/100 ml	10 g/100 ml	20 g/100 ml
WM-143-2-3b	78	39	53	25
Norbert	98	83	78	49
Fairfield	100	67	67	32
S77433	94	89	86	82

*
Grams of hull/ml of water

Germination of Norbert seed was not affected by the presence of WM-143-2-3b seed in any of the ratios attempted. However, germination was inhibited in the presence of an aqueous extract (10 gms hulls/100 ml water) of hulls of WM143-2-3b. The germination of the Norbert control was 99% and the Norbert treated 19%. The germination of the four parents was also reduced by aqueous extracts of S77431 hulls (Table 1). It is not clear why the hulless parent was less affected than the three hulled parents. Thus there appears to be water soluble constituents in the hulls which are capable of germination inhibition. This has been reported in barley, wheat and rice (Corbineau 1980, Dunwell 1981 and Mikkelsen and Sinah 1961). Further work is required to identify these constituents and their mode of action.

CONCLUSIONS

The dormancy present in the line WM143-2-3b has a low heritability but it should be possible to select segregates, from crosses with this line, which have a satisfactory level of dormancy. While there is a negative relationship between sprouting resistance and amylolytic enzyme activity segregants have been identified which appear to combine resistance with good quality. The factors responsible for the dormancy are located in the hull and further work is required to identify these factors and to elucidate their mode of action.

REFERENCES

Brookes, P.A. 1979. The significance of pre-harvest sprouting of barley in malting and brewing. Cer. Res. Comm 8: 29-38.

Corbineau, F. et Come, D. 1980. Quelques caracteristiques de la dormance du caryopse d'Orge (Hordeum vulgare L., variete Sonja). C.R. Acad. Sc. Paris 290: 547-550.

Dunwell, J.M. 1981. Dormancy and germination in embryos of Hordeum vulgare L. - Effect of dissection, incubation, temperature and hormone application. Ann. Bot. 48: 203-213.

Gordon, A.G. 1973. The rate of germination. Seed Ecology ed. W. Heydecker. Butterworths, London 578 pp.

Mikkelson, D.S. and Sinah, M.N. 1961. Germination inhibition in Oryza sativa and control by preplanting soaking treatments.

Crop Sci. 1: 332-335.
Mueller, H.W. 1964. About breeding of sprouting resistant cereals.
P. Br. Abstr. 34: 189.

Early Generation Mass Selection for Seed Dormancy in Barley

L. Reitan, *Voll Agricultural Research Station, Moholtan, Trondheim, Norway*

SUMMARY

Scandinavian spring barley production suffers from sprouting damage in rainy and wet autumns. Seed dormancy gives protection against this damage and is a wanted character in nordic feed-barley varieties. In breeding work it is of importance to start early generation selection for this character. Mass selection of dormant seed from bulk populations giving sub-populations with increased resistance to sprouting is described, and also a simplified method of standardizing water amount to the germination medium giving a coefficient of variation in water content at a level of CV=1.5%, and a coefficient of variation in germination at a level of CV=10%. Drying the samples and stopping germination makes it possible to handle large numbers of samples. Fairly good genetic advance by selection as early as in F_2 are shown.

INTRODUCTION

Sprouting damage in spring barley (<u>Hordeum vulgare</u> L. and <u>H. distichon</u> L.) is frequent in areas with heavy precipitation at harvest time in Scandinavia. Barley starch quality is less important when grown for animal feed, but sprouting in the ear damages the crop by causing problems in combine harvesting, by grain losses at harvest and as well makes storage difficult. Sprouted grain is easily destroyed by storage fungi.

Changed farming techniques with combine harvesting, heavy nitrogen manuring, and intensive crop production has led to enhanced danger of lodging and consequent sprouting risk for the crop if it remains unharvested for a period of time in bad weather.

Seed dormancy

Seed dormancy protects against sprouting damage, along with possible starch degradation. In the last twenty-five years, seed dormancy has been very important in Norwegian barley breeding programs (Strand 1970). Nearly all the Norwegian barley production is used as animal feed, and very little is used for brewing. In addition to real seed dormancy, resistance to sprouting in the ear may be a result of several factors and is dependent on ear shape and position, awning, thickness of husks, etc. Naked barley is often much more prone to sprouting than covered varieties. Cracking and rupture of pericarp/testa start the sprouting process. Thin-husked barley varieties are more easily ruptured and are, therefore, susceptible to sprouting (Brookes 1979).

Earliness and strong straw strength assist in preventing preharvest sprouting. Early varieties are often ready to be harvested before the usual wet weather conditions such as those which occur in mid-Norway. Earliness often is associated with a low degree of seed dormancy. Weak straw strength may lead to lodging, which in turn contributes to sprouting damage. The grains and awns are more easily wetted by rain, and the drying process occurs slowly.

Seed dormancy is the single factor which gives the best protection against sprouting. Seed dormancy is a complex and is the sum effect of several mechanisms acting together and in interaction with the milieu.

A barley variety for mid-Norway should have enough seed dormancy at autumn to protect against sprout damage in a lodged, over ripened crop for at least three weeks. Strand (1965) found that in barley seed dormancy was significantly influenced by variety, year, and germination temperature while the effect of harvest time was less certain. Interactions of variety x year and of variety x germination temperature were important, but variety x harvest time interaction was a less important cause of variation. Strand (1971) also showed that during storage of barley at low temperature (3°C) changes of seed dormancy occurred very slowly. After storage of barley at this temperature for four months, half the seed dormancy expressed as Dormancy Index, DI (Strand 1979), remained.

Breeding strategy

In a breeding program, it is important to start simple selections on important characters with a high degree of heritability as early as possible. It is also useful and effective to propagate breeding populations in bulk for three to four generations before single plant selections are made. Thus, some of the non-fixable heterosis-effects in early generations are removed, less heterozygosity occurs, action of natural selection are present and the breeder may get an impression of the value of the population.

Mass selection for seed dormancy in bulk barley populations can be done in early generations using the following suppositions:

i High seed dormancy in barley is usually a recessively inherited character.

ii Variation in seed dormancy is probably influenced by several genes.

iii Seed dormancy in barley varies with variety, year, germinating temperature, storage temperature and harvesting time, together with interactions between these main effects. Under standardized conditions, seed dormancy has a relatively high heritability. (Strand 1965, Buraas 1979).

iv Selection for seed dormancy as early as F_2 should be effective.

Mass selection for high resistance to sprouting in wheat may be carried out in alternate ways. Rain-simulating chambers are useful for detecting sprouting in ears of wheat (Svensson 1975, Derera et al. 1975, Gordon 1979). Weilenmann (1979) made selections in wheat by testing for Falling Number values with and without moisture chamber treatment (sprouting index) resulting in cultivars with improved sprouting resistance. Strand (1979) made barley selections in F_2 and later generations by germinating in sand, and Buraas (1979) also found selection for seed dormancy in barley F_2 generations to be effective.

Johansson (1977) showed that water sensitivity effects in wheat were more pronounced in moisture chambers because of excess water, and found a greater heritability of seed dormancy by selecting in petri dishes. Essery et al. (1964) found higher water sensitivity in dormant barley, and, therefore, it is necessary to standardize the amount of water in germination tests when selecting for high seed dormancy.

MATERIALS AND METHODS

Materials

Bulk populations of barley in F_2 - F_4 were grown in dense plots in 1979, harvested in bulk at normal combine harvesting time, dried, and stored in a cool place (not over 10°C) for three months. Samples of each bulk population were placed into a seed dormancy selection sequence as described below. Table 1. shows a few of the populations under investigation.

Methods

Batches of 300 folded paper towels were used as germination paper and placed vertically in a suitable polyethylene box containing 3000 mL of fresh tap water. These were left for two to three hours for equilibrium to occur such that each paper contained 10 mL of water. This method was developed by Reitan (1977).

Table 1. Pedigree and generation for populations under selection.

Cross no.	Generation 1979	Pedigree
H76018	F_4	Yrjar//Forus/EddaII (715-64)
H77803	F_3	Jarle/Otra (1128-66)//Tunga
H77162	F_3	Agneta/Yrjar
H78139	F_2	Agneta/H306-8
H78140	F_2	H306-8/Yrjar
H78141	F_2	H306-8/Tunga

Variation in water soaking was very small with a coefficient of variation (CV) of about 1.5%. Thus even amounts of water are distributed to each sample of grain. One hundred grains were placed on each of 30 papers for a total count of 3000. The paper towels were folded, ten samples put together in a polyethylene bag and all the bags placed in a germination chamber with constant temperature of $15°C \pm .5$ for seven days. The samples were then taken out of the plastic bags and placed in a drying chamber at 28-30°C for one day in order to stop germination. Earlier investigations (Reitan 1977, 1979) have shown that the method gives a coefficient of variation for seed germination of approximately 10%.

After drying, the seeds were removed from the papers and each population bulked, de-awned and cleaned in a seed cleaner. The sprouted seeds lost viability by this procedure. In the following spring each selected sub-population was sown in a field plot.

From some of these selected sub-populations a limited number of ears were selected in 1980. Single-head rows were grown in 1981 and new selections on other important agronomic characters were made. Selected strains were tested for seed dormancy index (DI) together with the unselected bulk populations and the dormancy-selected sub-populations during the following winter.

RESULTS AND DISCUSSION

Results of seed dormancy investigations for six bulk populations and corresponding selected sub-populations in 1980 and 1981 are presented in Figure 1.

One of the parents in the populations, the six-row Svaløf variety Agneta, has very high resistance to sprouting and is one of the parents of H77162 and H78139. The highest genetic advance also occurs in these populations. In both years the seed dormancy level in the selected sub-populations were significantly higher than for the original population, and gave a fairly good prediction of expected sprouting resistance when selecting ears for other agronomic important characters later on.

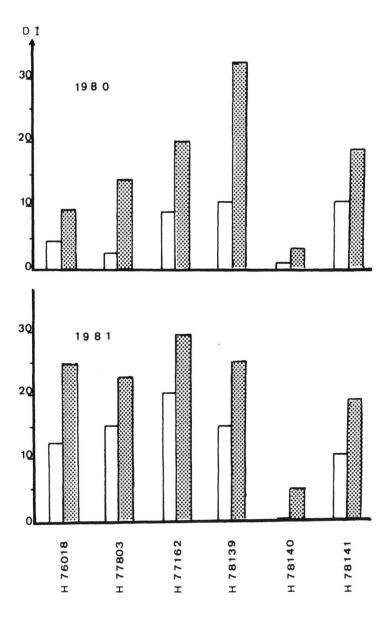

Fig. 1. Bulk population means for seed dormancy index (DI)

☐ unselected pop. ▓ selected sub.pop.

Seed dormancy indexes for unselected and selected sub-populations, selection intensity (%) in 1979, and mean DI in 1981 of selected strains from the sub-populations are presented in Table 2. The selected strains in the middle are more dormant than the cor-

Table 2. Results from one cycle bulk selection for seed dormancy. DI=Dormancy Index.

| Cross no. | Sel. | DI 1980 | | DI 1981 | | Selection in lines | | |
| | | unsel. | sel. | unsel. | sel. | N | $\overline{DI}$ | sd. |
	%							
H76018	51	4.7	9.0	12.7	25.3	28	48.1	35.4
H77803	26	2.7	14.0	15.3	23.0	16	24.0	17.1
H77162	16	8.0	20.3	21.0	29.0	17	41.6	27.2

responding sub-population means. This is probably due to less lodging and more tillering in singlerows. By growing on larger plots in 1982 an answer will be found as to whether these positive effects are significant. In this mass selection program, the selection intensity for seed dormancy was not too high (from 16 to 50%). In spite of this, the genetic advance is significant.

The method described in this paper has many advantages. It requires minimal labour, is inexpensive, clean to work with, and needs little germinating chamber space. The method is very convenient because germination is stopped by heat drying (not warmer than 35°C). Storing of the dried samples allows analyses at a time other than the busy harvest.

By looking at germination development after the fifth day, one can regulate selection intensity by interrupting the process earlier or later than on the seventh day. For low seed dormancy levels, a raise in germinating temperature up to some 20°C is recommended.

A great variation in earliness within populations may cause a corresponding negative response on earliness when selecting for seed dormancy. Such responses are found in wheat using mass selection (Svensson 1975). One way to overcome such a problem is to make sub-populations selected for classes of earliness.

REFERENCES

Brookes, P.A. 1979. The significance of pre-harvest sprouting of barley in malting and brewing. Cereal Res. Commun. 8:29-38.

Buraas, T. 1979. Analyse av avkom etter konvergente krysninger i bygg. Dr. scient. Theses, Norges Landbrukshogskøle.

Derera, F.N., McMaster, G.J., and Balaam, L.M. 1975. Pre-harvest sprouting resistance and associated components in 12 wheat cultivars. Cereal Res. Commun. 4:173-180.

Essery, R.E., Kirsop, B.H., and Pollock, J.R.A. 1964. Studies in barley and malt. I: Effects of water on germination. J. Inst. Brew. 60:473-481.

Gordon, I.L. 1979. Heritability of grain development trait associated with sprouting damage in wheat inbreds. Cereal Res. Commun. 8:185-192.

Johansson, N.E. 1977. Studier av fältgroingsresistensen hos vete - bestemmingsmetoder - sortsdifferenser och nedarvingsförhållanden. Thesis, Lantbrukshögskolan, Uppsala, Sweden.

Reitan, L. 1977. Kvalitetsmål og analysemetodikk i varresistenssammenheng. Seminar, NLH des. 1977. NLVF-publ.

Reitan, L. 1979. Genetical aspects of seed dormancy in wheat related to seed coat colour in an 8 x 8 dialele cross. Cereal Res. Commun. 8:275-282.

Reitan, L. 1979. Undersøkelser over det genetiske grunnlag for spiretreghet hos hvete. Dr. scient. Thesis, Norges Landbrukshøgskole.

Strand, E. 1965. Studies on seed dormancy in barley. Meld. Nor. Lundbrukshoegsk 49(7):1-22.

Strand, E. 1970. Spiretreghet hos sorter og foredlings-materiale av hosthvete, varhvete, bygg og havre. Meld. Nor. Lundbrukshoegsk 49(2):1-23.

Strand, E. 1979. A seed dormancy index for selection of cereal cultivars resistant to pre-harvest sprouting. Cereal Res. Commun. 8:219-225.

Svensson, G. 1975. A selection method to test the sprouting resistance in wheat. Cereal Res. Commun. 4:263-266.

Weilenmann, F. 1979. Plant breeding aspects of sprouting resistance with the selection technique using sprouting index. Cereal Res. Commun. 8:209-218.

Recombining Dormancy from RL 4137 with White Seed Color

R. M. De Pauw and T. N. McCaig,
Research Station, Agriculture Canada,
Swift Current, Saskatchewan, Canada

SUMMARY

RL 4137, a spring wheat genotype with a long, stable dormancy period and three genes for red seed color, was hybridized with 7722, a white-seeded wheat. In both the F3 and F5 generations a positive relationship existed between red seed color and sprouting resistance (SR). The mean number of heads sprouted per sample for all the red-seeded progeny was less than that for the red-seeded control cultivars in both the F3 and F5.

The six white-seeded F3 lines exhibited a range in sprouting response from susceptible to as resistant as some red-seeded control cultivars. The mean number of heads sprouted per sample for two white-seeded F4 families was intermediate to both the red-seeded and white-seeded controls at both T_0 and T_{10}. Some white-seeded F4 lines had lower sprouting at T_{10} than the red-seeded controls Pitic 62, Neepawa, and Glenlea.

The percent dormancy of six white-seeded F5 families derived from F3-lines was greater than the mid-parent value. There were significant differences among the F5 families for mean percent dormancy.

The results indicate that some of the dormancy of RL 4137 has been transferred to the white-seeded progeny. The evidence suggests that RL 4137 has a genetic mechanism for SR associated with red seed color and one or more mechanisms not associated with seed color.

INTRODUCTION

The breadmaking process is adversely affected by high levels of alpha-amylase caused by de novo synthesis during the germination process. Increased consideration has been given to wheat production research and development techniques to provide wheat of low alpha-amylase

activity for commercial processing. Briggle (1980) reported a harvesting system used by farmers in Michigan
that has reduced sprout damage in wheat from a serious
problem to one of minor importance. Another approach to
reduce preharvest sprouting has been to breed cultivars
with an increased dormancy period.

The main attribute of the spring wheat cultivars
Columbus and Leader is a long dormancy period (Campbell
and Czarnecki 1981; De Pauw et al. 1982). The dormancy
of both of these cultivars appears to derive from the
Brazilian cultivar Frontana. The original reason for
using Frontana was to transfer leaf rust resistance from
Frontana into North American cultivars. Examples of
cultivars produced from such programs are Chris (Fron-
tana/3*Thatcher/3/Kenya 58/Newthatch//2*Thatcher) and
RL 4137 (Frontana/3/McMurachy/Exchange//2*Redman/4/
Thatcher*6/Kenya Farmer). Both Chris and RL 4137 also
have very good dormancy, which might be a result of
linkage with gene(s) for leaf rust resistance or merely
a fortuitous event.

The dormancy of RL 4137 was transferred by back-
crossing to produce Columbus (Campbell and Czarnecki
1981). A two-way cross followed by selection for both
stem solidness and low alpha-amylase activity was sufficient to recombine the dormancy of Chris with the solid
stem characteristic of Fortuna (De Pauw et al. 1982).
The long stable dormancy of Frontana appears to have
been easily transferred to red-seeded wheats.

The objective of this study was to determine if the
dormancy of RL 4137 could be recombined with white seed
color.

MATERIALS AND METHODS

Experiment I. Variability for Sprouting Resistance in a
Cross of RL 4137 and a White-Seeded Line

RL 4137, a spring wheat with a long stable dormancy
and three genes for red seed color, was hybridized with
7722, a white wheat. F3 lines derived from single F2
plants, parents, and control cultivars were grown in 3-m
long plots. Ten random heads were collected from each
plot. Two weeks later another set of ten heads was collected from the controls and six white-seeded lines.
The head bundles were subjected to 50 to 60 mm of simulated rain over a 3-hr period in a controlled environment cabinet. After six days at 18°C and 100% relative
humidity, the number of heads with visible sprouts per
sample, the percentage of kernels germinated, and Hagberg Falling Number were determined.

In 1981, F5 families that derived from selected F3
lines and control cultivars were grown in 3-m long

plots. The progeny test consisted of collecting ten random heads from each plot, subjecting them to a simulated rain, and counting the number of heads with visible sprouts per sample.

As part of the F5 progeny testing program, in 1981, measurements of variability due to environment and experimental technique were conducted. Samples of ten random heads were collected from eleven control cultivars that had been sown on four dates between May 4 and 21. The head samples were assigned random positions in the rain simulation chamber on six different cycles of testing.

Experiment II. Variability for Sprouting Resistance Within Two White-Seeded F5 Families

Control cultivars, parents, and 101 F4 lines that derived from two white-seeded F3 lines were grown in 3-m long plots. A sample of ten random heads was collected from each plot at about 20% moisture (T_0) and again 10 days later (T_{10}). Both sets of head samples were subjected to a simulated rain. After seven days the samples were removed and the number of heads with visible sprouts counted.

Experiment III. Variability for Sprouting Resistance Within Six White-Seeded F5 Families

A bulk of F5 seed was produced from each of the six white-seeded F3 lines from the cross 7722/RL 4137. Ten to forty plants of each F5 bulk and parents were grown in a controlled environment room under a 16-hr photoperiod at 20°C and a 8-hr nyctoperiod at 15°C. Spikes were tagged at anthesis for identification. Each spike was harvested 70 days after anthesis.

Seeds from a single spike were placed on filter paper in a petri dish to which was added 5 ml water. The seeds were placed in a dark germination cabinet at 22°C. After four days the number of seeds which had germinated were counted. Those seeds that had not germinated were placed in a 0°C environment for two days and then put back into the germinator for another two days. Dormancy was calculated as a percentage of those seeds which germinated after a low-temperature treatment of the total seeds germinated.

RESULTS AND DISCUSSION

Experiment I. Variability for Sprouting Resistance Among the Progeny of 7722/RL 4137

Mean squares for arcsine number of heads sprouted

per sample for white vs. red-seeded F3 lines were significant. The mean number of heads sprouted per red-seeded F3 line was 1.04 (Table 1). Average sprouting for the six white-seeded lines and also for the four red-seeded control cultivars was 3.7 heads for the first sampling date. Averaged over both sampling dates, the F3 lines QV and QX exhibited lower sprouting than both the other white-seeded lines and the moderately sprouting resistant white wheat Kenya 321. The lines QV and QX appeared to have more sprouting resistance than the red-seeded control cultivars NB 320, Glenlea, and Neepawa. Lines QW and QY seemed to have a short dormancy period.

Table 1.
Number of heads with visible sprouts in a sample of ten heads, percentage of kernels sprouted, and Hagberg Falling Number for control cultivars and six white-seeded F3 lines sampled at two intervals

Genotype	Seed color	No. of heads sprouted		% kernels sprouted		H.F.N. value	
		Time		Time		Time	
		0	+14 days	0	+14 days	0	+14 days
NB 320	light red	7		81		62	
Glenlea	red	6		73		62	
Neepawa	red	2		76		86	
RL 4137	red	0	1	1	2	385	382
Kenya 321	white	0	10	26	83	216	72
F3 lines							
QU	white	8	10	96	92	62	62
QV	white	2	2	15		180	
QW	white	3	9		83		62
QX	white	0	4	50		158	
QY	white	0	9	59	74	195	62
QZ	white	9	8	92	85	62	62
Mean of 310 red-seeded lines		1.04					

H.F.N. Hagberg Falling Number

In 1981 variability in sprouting response due to environment and experimental technique was detected but was present at an acceptable level. For the thirteen 10-head sample of RL 4137, no heads with visible sprouts were observed (Table 2). Pitic 62 had the highest mean sprouting per sample and the largest standard error of the mean among the red-seeded cultivars.

Table 2.
Mean, minimum and maximum number of heads sprouted for samples of ten heads per genotype after treatment in rain simulator

Genotype	Color	No. of samples	Mean No. of heads sprouted/ sample	Min.	Max.	Std. error of mean
RL 4137	R	13	0	0	0	0
NB 320	LR	22	0.95	0	3	0.21
Glenlea	R	6	1.38	0	7	0.48
Neepawa	R	12	2.00	0	5	0.46
Pitic 62	R	6	5.50	1	9	1.48
7722	W	4	9.00	8	10	0.41
Kenya 321	W	14	9.21	8	10	0.21
NB 112	W	4	9.75	9	10	0.25
NB 402	W	7	9.86	9	10	0.14
Norquay	W	7	10.0	10	10	0
Fielder	W	6	10.0	10 ·	10	0

length of the coleoptile reached a maximum of 25 mm. Kenya 321 germinated neither rapidly nor uniformly. The percentage of kernels with visible sprouts ranged from 5 to 25 percent, and the coleoptile, if visible, was short (<5 mm).

In 1981 the red-seeded cultivars expressed more resistance to sprouting than white-seeded cultivars (Tables 1 and 2). There were differences, however, in degree of sprouting within both seed color classes.

Mean squares for arcsine number of heads sprouted for white vs. red-seeded F5 lines were significant. These results corroborate the general positive relationship between red seed color and dormancy (Gfeller and Svejda 1960; Khan and Strand 1977).

The average number of heads sprouted per line in both the F3 and F5 generations was less than the average for the red-seeded controls (Tables 1 and 3). Also, the average sprouting for the white-seeded lines tended to be less than the white-seeded controls. This suggests that the dormancy of RL 4137 may be controlled by more than one mechanism, one of which may be associated with seed color and one or more may not be associated with seed color. Evidence for a genetic mechanism(s) not associated with seed color could be obtained by examination of white-seeded progeny.

256

Table 3.
Mean number of heads sprouted/sample of F5 lines, parents, and red-seeded control cultivars

Genotype	n	Mean No. of heads sprouted/sample
White-seeded F5 lines	15	9.07
F5 lines segregating for color	41	2.24
Red-seeded F5 lines	80	0.61
Red-seeded controls	4	1.25
RL 4137	2	0
7722	2	10.0
White-seeded controls	4	10.0

Experiment II. Variability for Sprouting Resistance Within Two White-Seeded F4 Families

The red-seeded control cultivars had the lowest mean number of heads sprouted per sample for both sampling times, while the white-seeded controls had the highest mean (Table 4). The mean number of heads sprouted for the two white-seeded families was significantly different from both the red-seeded controls and the white-seeded controls at both sampling times.

Table 4.
Mean and standard error of the mean for number of heads sprouted per sample of 10 heads per sample

	No. of genotypes	Time of sampling	
		0	+10
Red-seeded cultivars	11	0.7 ± 0.24	4.3 ± 1.03
White-seeded cultivars	5	9.4 ± 0.24	9.8 ± 0.20
White family - QV	20	6.7 ± 0.57	6.7 ± 0.59
White family - QX	81	6.8 ± 0.26	8.1 ± 0.19

The red-seeded controls differed in the length of their dormancy period (Table 5). At T_0 the range in sprouting per sample was from 0 to 2. At T_{10} the range was from 0 to 10 heads sprouted per sample. At T_{10} Pitic 62 had the most heads sprouted and RL 4137 had the least. The long, stable dormancy of RL 4137 and the unstable dormancy of Pitic 62 are supported by the data in Table 2. Both samples of Kenya 321 had 9 heads sprouted at the initial sampling date and averaged 9.5 heads sprouted at the second sampling date.

The distribution of the number of heads sprouted

per sample of the F4 lines of the white-seeded family QV spanned the range from 0 to 10 at both T_0 and T_{10} (Table 5). Similarly, the white-seeded family QX had a broad range. Some of these white-seeded lines had a longer dormancy period than the red-seeded cultivar Pitic 62 and Kenya 321, which is considered to be one of the more sprouting resistant white wheats available (Derera et al. 1977).

Table 5.
Distribution of the number of heads sprouted per sample of control cultivars and two white-seeded F4 families

Time of sampling	Wheat class	No. of heads sprouted/sample			
		0-2	3-5	6-8	9-10
T_0	Red controls	11	0	0	0
	White controls	0	0	0	5
	QV family	1	4	9	6
	QX family	5	18	38	20
T_{10}	Red controls	4	3	3	1
	White controls	0	0	0	5
	QV family	2	6	6	6
	QX family	0	9	31	41

Experiment III. Variability for Dormancy Within Six White-Seeded F5 Families

The mean squares for arcsine percent dormancy of genotypes were significant as were all single degree of freedom comparisons except among the dormant white-seeded familites QW, QV, and QX (Table 6).

Table 6.
Analysis of variance for the arcsine of percentage dormancy for parents and six white-seeded F5 families

Source	df	ms
Genotypes	7	2.85**
P_1 vs. P_2	1	8.23**
P_1 + P_2 vs. Progeny	1	1.26**
Dormant vs. Nondormant progeny	1	6.48**
QW vs. QV + QX	1	0.07
QV vs. QX	1	0.16
QU vs. QY + QZ	1	1.00**
QY vs. QZ	1	1.47**
Error	607	0.08

Since a maximum differential in percentage dormancy was obtained between the sprouting resistant parent RL 4137 and the susceptible parent 7722 at 70 days after anthesis (Table 7), each tagged head of parent and progeny was harvested sequentially 70 days after flowering. The means of the six families ranged from 64 to 96 percent dormancy, which was significantly greater than the mean of 7722. The range for percentage dormancy within all families except QW was large.

Table 7.
Mean, minimum, and maximum percentage dormancy for parents and six white-seeded F5 families

Genotype	No. of plants	No. of heads	% dormancy		
			Mean	Min.	Max.
RL 4137	10	40	99.7	98.3	100.0
7722	10	42	35.9	24.3	52.3
7915-QW	10	43	96.4	83.3	99.4
-QV	40	170	92.9	39.8	99.3
-QX	39	179	87.6	45.1	99.2
-QU	10	47	73.2	35.2	96.8
-QY	10	54	66.9	33.5	94.2
-QZ	10	40	64.2	1.2	94.6

The variability for dormancy within the white-seeded progeny of 7722/RL 4137 spans the range from the white-seeded parent 7722 to red-seeded controls like Pitic 62, Neepawa, Glenlea, and NB 320. This suggests that some of the dormancy of RL 4137 was transferred to the white-seeded progeny. None of the white-seeded progeny, however, equalled the dormancy of RL 4137. The evidence supports the hypothesis that RL 4137 has more than one mechanism controlling dormancy.

REFERENCES

Briggle, L.W. 1980. Pre-harvest sprout damage in wheat in the U.S. Cereal Res. Commun. 8:245-250.

Campbell, A.B., and Czarnecki, E.M. 1981. Columbus hard red spring wheat. Can. J. Plant Sci. 61:147-148.

De Pauw, R.M., McBean, D.S., Buzinski, S.R., Townley-Smith, T.F., Clarke, J.M., and McCaig, T.N.]982. Leader hard red spring wheat. Can. J. Plant Sci. 62:231-232.

Derera, N.F., Bhatt, G.M., and McMaster, G.J. 1977. On the problem of pre-harvest sprouting of wheat. Euphytica 26:299-308.

Gfeller, F., and Svejda, F. 1960. Inheritance of post-harvest seed dormancy and kernel colour in spring wheat lines. Can. J. Plant Sci. 40:1-6.

Khan, F.N., and Strand, E.A. 1977. Investigations into the genetics of kernel color and dormancy in wheat (Triticum aestivum L.). Meld. Nor. Landbrukshogsk. 56:1-12.

Effects of Temperature and Rainfall on Seed Dormancy of Small Grain Cultivars

Erling Strand, *Agricultural University of Norway*

INTRODUCTION

A number of investigations have been carried out in order to study the effects of temperature and rainfall on the intensity and duration of seed dormancy of cereals in field prior to harvest. Belderok (1965 and 1968) found a close connection between the temperature sum above 12.5°C during the dough ripe stage and the length of the dormancy period. Based on these findings he proposed a scheme for warning wheat growers when wheat fields became susceptible to ear sprouting. Olsson and Mattsson (1976), Lallukka (1976), Grahl and Schrödter (1973), Mitchell et al (1980) etc. did not find an adequately close correlation between weather conditions and the duration of the seed dormancy to provide a warning of sprouting in farmers' fields.

At the Department of Farm Crops, investigations of the effects of weather conditions on pre-harvest ear sprouting and other quality damage of small grain cereals have been carried out since 1951. Since 1959 routine tests for seed dormancy have been used for selection of cultivars and breeding material to improve resistance against pre-harvest ear sprouting.

The present report is an attempt to determine more closely the effects of temperature and rainfall on the intensity and duration of seed dormancy in field prior to harvest, and to investigate the reasons for the varying results obtained in earlier studies.

MATERIAL AND METHODS

In this study three cultivars were grown over a period of 16 years: the six-rowed barley cv. Lise, the two-rowed barley cv. Møyjar and the spring wheat cv. Runar. The oat cv. Mustang was grown for 11 years. For a part of the study, other cultivars are also included. All cultivars are well adapted and have been grown widely in this area of Norway.

The cultivars were grown in 10 m^2 plots with 13.3 cm row spacing and a seed rate of 200 kg per ha. In all years the plots were sampled at two stages of maturity for determination of seed dormancy, ten days post yellow ripeness (YR) and 30 days post YR. In some years at the beginning of the period the plots were sampled at four maturity stages; YR, and 10, 20 and 30 days post YR. The maturity stage "yellow ripeness" was defined as 38% grain moisture, determined by frequent sampling for moisture content as the plots visually approached yellow ripeness.

Seed dormancy was determined by germination tests of 200 kernels in moist sand at 12°C and 20°C for 10 days following Official Seed Testing Procedures. Dormancy Index (DI) was calculated as the mean percentage of dormant seeds obtained at the two germinating temperatures. The results at 12°C are given double weight. DI was used to cover possible cultivar maturity differences and cultivar x germination temperature interactions and to obtain more representative figures for the dormancy level of the cultivars at harvesttime.

In order to study the effects of temperature and rainfall on seed dormancy at different stages of seed development, the period from anthesis to maturity was divided into the following periods:

1. 0-10 days post anthesis
2. 20-10 days prior to YR
3. 10-0 days prior to YR
4. 0-10 days post YR
5. 10-30 days post YR.

For each of these periods the following climatic parametres were calculated.

1. Daily mean temperature
2. Mean temperature of the six hottest hours of the day
3. Mean daily max. temperature
4. Sum of rainfall
5. Global radiation.

The effects of these parametres on the intensity of seed dormancy at different maturity stages were investigated by applying partial correlation and multiple correlation techniques. Because of the limited number of degrees of freedom (maximum 16 years for each cultivar) compared to the high number of independent variables, a maximum of 10 independent variables were included in the analyses of each cultivar.

The three temperature parametres are strongly inter-correlated and also strongly correlated with global radiation. Preliminary analyses showed that the use of maximum temperature or the temperature of the six hottest hours of the day did not improve the correlations with seed dormancy in relation to the daily mean temperature.

Neither did the global radiation give any higher correlations with seed dormancy than the daily mean temperature. In the final analyses, therefore, the daily mean temperature and the sum of rainfall were used for each of the five 10 day periods.

RESULTS

For information on the seed dormancy of the cultivars some key data are listed in Table 1. The lowest, the means and the highest values of seed dormany are recorded during the 16 year period at the two harvest stages for cv. Lise. During that period, the percent dormant seeds harvested 10 days post yellow ripeness for the cv. Lise varied from 29 to 93%, and similar results were obtained for the other cultivars. Table 1 also shows that the mean dormancy is reduced from the first to the second harvest. The lowest, mean and highest temperature value and rainfall for each 10 day period are also presented in Table 1.

The effects of temperature and rainfall on the intensity of seed dormancy were calculated and are given in Table 2. The effect of each variable was estimated by step-wise elimination of the different independent variables. For cv. Lise, harvested 10 days post yellow ripeness, 1.7% of the variation in seed dormancy can be explained by the variation in temperature during the 10 day period of post anthesis. Variation in rainfall during the same period contributed 38.4% of the variation. In sum, the eight independent variables were responsible for 65.3% of the variation in seed dormancy corresponding to a multiple correlation coefficient of 0.80**. The other multiple correlation coefficients are also significant. The effect of temperature and rainfall in the different 10 day periods, however, varies greatly between cultivars and between maturity stages. Surprisingly, the amount of rainfall during the first 10 days of post anthesis seems to have the strongest and the most consistent effect on the development of seed dormancy. The strong effects of temperature during the 20 days prior to harvest at the latest maturity stage are more in accord with earlier findings, i.e., that dormancy is reduced at higher temperatures.

The irregular effects of environmental factors on seed dormancy indicate strongly that the real factors affecting seed dormancy are not recorded or that the estimates of seed dormancy were not sufficiently precise.

It was shown in Table 1 that there were large year to year differences in seed dormancy. Analyses of variance showed highly significant differences in seed dormancy between cultivars, years, time of harvest and germinating temperatures. The following interactions were also highly significant:

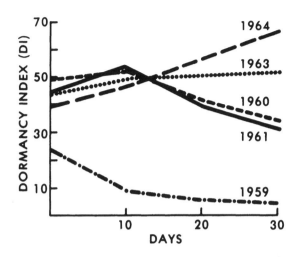

Figure 1. Intensity of seed dormancy in different years.

Table 1. Lowest, mean and highest values of two dependent and 8 or
10 independent variables. Period of 16 years for the cv.
Lise.

	Lise		
	Low	Mean	High
Dependent variables			
1. DI 10 days post YR	29	49.6	93
2. DI 30 days post YR	9	29.7	66
Independent variables			
1. Temp. 0–10 days post anthesis	13.8	16.7	20.5
2. 20–10 days prior to YR	13.1	15.6	18.5
3. 10–0 days prior to YR	14.2	16.4	18.8
4. 0–10 days post YR	12.3	15.4	19.4
5. 10–30 days post YR	12.3	14.7	19.9
6. Rainfall: 0–10 days post anthesis	0	22.9	68
7. 20–10 days prior YR	0	31.2	95
8. 10–0 days prior YR	0	22.1	53
9. 1–10 days post YR	0	16.7	43
10. 10–30 days post YR	1	16.8[1]	103

[1] Rainfall, mean 10 day period.

Cultivar x year
Cultivar x time of harvest
Cultivar x germinating temperature
Harvest time x germinating temperature.

The cultivar differences in seed dormancy between years are no doubt due to variation in the climatic conditions. The highly significant interactions listed above must indirectly also be due to variation in climate. Cultivar interactions with years and time of harvest may explain the apparently inconsistent results in Table 2. Another reason for inconsistent results are to be found in seed dormancy estimates.

Figure 1 shows seed dormancy as mean values of a number of wheat, barley and oat cultivars recorded at YR, and 10, 20 and 30 days later for five different years. Seed dormancy attains maximum values at different stages of maturity during different years. At predetermined times of sampling for seed dormancy, as in this investigation, the time of sampling may not be at the time of maximum seed dormancy. In such cases, the values are not correct estimates of seed dormancy, yet these values are used to determine correlations of seed dormancy with climatic factors.

In the warm season of 1959, seed dormancy attained maximum values prior to or at YR. At 10 and 30 days post YR, dormancy was reduced strongly. In 1960 and 1961, dormancy was normally developed. Maximum dormancy was reached 10 days post YR but was reduced by approximately 1.0% per day in the following 10 day periods. In the cool seasons of 1963 and 1964, dormancy increased at least until 30 days post YR.

The data accumulated for 16 years indicate that the most common pattern for development of seed dormancy is as found in 1960 and 1961. Developmental patterns for seed dormancy in 1959, 1964 and partly in 1963 are exceptions. In the way the sampling for dormancy was carried out, however, the results obtained in such years may reduce the correlations between dormancy and climatic factors, and, in some cases, also give misleading results.

A scheme for warning of susceptibility to pre-harvest ear sprouting based on the temperature sum (Belderok 1965) may be satisfactory if climatic conditions are stable from year to year, i.e., dormancy level and the time of maximum dormancy show small variations from year to year. In cases of greater annual variation in climatic conditions, as in Norway, a scheme for warning of susceptibility to pre-harvest ear sprouting should be based on the following information.

1. The value of maximum seed dormancy
2. The time that seed dormancy is at maximum
3. The rate of disappearance of seed dormancy.

Table 2. Percent variation in dormancy (DI) due to variation in temperature and rainfall in different 10 day periods. The + or - signs indicate positive and negative correlation, respectively.

	Cultivar			
	Lise 16 years	Moyjar 16 years	Runar 16 years	Mustang 11 years
Seeds harvested 10 days post yellow ripeness				
Temp:				
0-10 days post anthesis	-1.7	-11.8	6.2	+17.1
10-20 days prior to YR	+1.7	+2.4	+1.1	+0.0
0-20 days prior to YR	-5.4	+2.7	-9.7	-7.3
0-10 days post YR	-2.1	-1.4	-3.4	+13.3
Rainfall:				
0-10 days post anthesis	+38.4	+11.7	+17.6	+7.7
10-20 days prior to YR	+0.3	+0.1	-6.5	-2.2
10-0 days prior to YR	+9.7	+9.7	-14.6	-4.9
0-10 days post YR	+6.0	0.0	0.0	-31.0
$(R^2.100)$	65.3	39.8	59.1	83.5
R	0.80	0.63	0.77	0.91
Seed harvested 30 days post yellow ripeness				
Temp:				
0-10 days post anthesis	-4.1	-2.2	+0.9	+6.9
20-10 days prior to YR	-4.9	-1.0	+0.4	-0.2
10-0 days prior to YR	-6.7	-5.7	+0.8	-1.5
0-10 days post YR	-4.3	-6.3	-0.4	-56.5
10-30 days post YR	-18.3	-52.4	-14.4	-17.6
Rainfall:				
0-10 days post anthesis	+5.6	+0.7	-3.4	+0.1
20-10 days prior to YR	+7.0	-0.1	-11.8	-0.3
10-0 days prior to YR	+29.3	+11.9	-0.9	-1.4
0-10 days post YR	+2.8	-1.5	-8.4	-10.1
10-30 days post YR	-0.2	-1.6	-5.5	+5.4
$(R^2.100)$	83.1	83.5	46.9	100.0
R	0.91	0.91	0.68	1.00

This information can be obtained by frequent sampling of fields (preferably twice a week) from YR and onwards. When the level of dormancy and date of maximum development of dormancy are determined, dormancy from then on is reduced by a certain rate, approximately 1.0% per day depending on temperature. Because seed dormancy determinations take 7-10 days, it is important to begin early in order to obtain an estimate of both season and cultivar seed dormancy levels.

The level of seed dormancy required for satisfactory high resistance against pre-harvest ear sprouting depends on many factors or conditions. Most important are the prevailing weather conditions, the expected length of periods of delayed harvest, the chance of quality damage that farmers are willing to take, etc.

REFERENCES

Belderok, B. 1965. Einfluss der Witterung vor der Ernte auf die Keimruhedauer und die Auswuchsneigung des Weizens. Acker-Pflanzenbau 122:297-313.

Belderok, B. 1968. Seed dormancy problems in cereals. Field Crop Abstr. 21:203-211.

Grahl, A. and Schrödter, M. 1973. Auswuchsvorhersage bei Weizen. Acker-Pflanzenbau 137:233-237.

Lallukka, V. 1976. The effect of the temperature during the period prior to opening or sprouting in the ear in rye and wheat varieties grown in Finland. Cereal Res. Commun. 4:83-91.

Mitchell, B., Black, M. and Chapman, J. 1980. Observation on the validity of predictive methods with regard to sprouting. Cereal Res. Commun. 8:239-244.

Olsson, B. and Mattsson, B. 1976. Seed dormancy in wheat under different weather conditions. Cereal Res. Commun. 4:181-185.

Pre-Harvest Sprouting—The South African Situation for Seed Dormancy in Barley

G. F. Marais and W.J.G. Kruis,
Small Grain Centre, Bethlehem, South Africa

SUMMARY

A summary of breeding and research events following the implementation of a pre-harvest sprouting screening program in 1979 is presented. Production areas subject to conditions favoring sprouting and problems encountered in breeding wheat for these areas are outlined. Relatively warm ripening conditions present a major challenge for breeders wishing to safeguard varieties from weather damage. There is a need for identification and utilization of components of sprouting resistance, apart from dormancy. High alpha-amylase activities in specific varieties not subjected to rainy pre-harvest conditions are presented.

INTRODUCTION

During 1970 to 1980, the average South African wheat crop amounted to 1.72 million metric tons per year (Wheat Board, 1981). The main production areas are shown in Figure 1. Areas traditionally encountering pre-harvest sprouting include the winter rainfall areas of the South Western Cape, namely the Ruens and Swartland. Spring wheats which contribute an average of 25.2 percent of the total crop, are harvested in this area during early summer. In 1974 and 1980, wheat in these areas suffered severe sprouting damage. An estimate of crop losses in 1980 assessed by the falling number method was 25% of the total crop. This amounts to approximately 5.5% of the national crop. The North Eastern Orange Free State (Fig. 1.) is a summer rainfall region and is a second problem area. The crop consists mainly of winter and intermediate wheats that are harvested during early to mid-summer. About 58.7% of the national crop is grown in this area (Wheat Board, 1980). Significant sprouting losses occurred in this area in 1973 and 1975. Estimated losses were 7.2% and 4.7% of the national crop, respectively (Laubscher, 1982). These estimates were based on visible sprouting and the real damage may have been higher. Smaller, though appreciable, losses also occurred on various occasions in these areas - Orange Free State in 1974 and the South Western Cape in 1976 and 1978 (Wheat Board, 1971-1981).

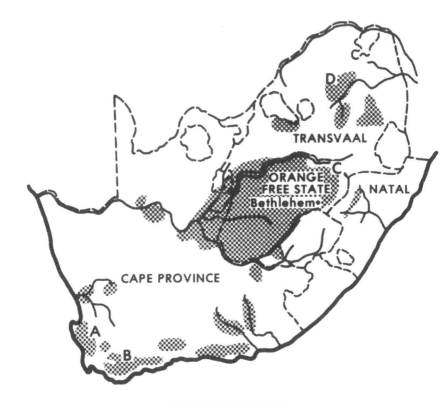

MATERIAL AND METHODS

Standard dormancy test

Dormancy was measured by means of germination tests done at 0, 10, 20 and 30 days after harvest ripeness (14% moisture). Germinations were conducted at 20°C over a seven day period using Whatman No. 1 filter paper.

Germination – temperature method of dormancy testing

The procedure described by Noll and Czarnecki (1979) was used.

Chaff inhibition

The presence of chaff inhibitors was tested using the method of Derera et al. (1977).

Germplasm used in the factorial experiment (temperature study)

One cultivar, Inia, and three breeding lines, M12/19, M1/18 and SST 22, were used in this experiment. The three breeding lines had the following pedigrees:

M12/19: Bb/Tob//Cno/Pato
M1/18: Cno "S" - Gallo (Vcm X Cno "S" - 7C/Kal -Bb)
SST 22: Flameks*3/Wernstein

Alpha-amylase determination

The nephelometric method described by Campbell (1979) was used.

RESULTS AND DISCUSSION

Breeding for sprouting resistance

A selection and research program aimed at combating susceptibility to pre-harvest sprouting damage was initiated at Bethlehem
(Fig. 1). The screening program involves three stages. F_2 and F_3
selections are screened according to the procedure described by
Svensson (1975). When 50 percent germination is attained, the
material is dried, threshed and subjected to visual screening. In
the case of F_4 and F_5 families, the procedure consists of cutting 15
ears at the same developmental stage as soon as 14% moisture is
attained. Seven days later the ears are placed inside a rain simulation cabinet for a further seven days. Visibly sprouted ears are
counted daily. Finally, advanced lines are evaluated in a rain simulation cabinet using a procedure similar to that described by
McMaster and Derera (1975). Sprouting resistance is defined as the
time interval (days) between harvest ripeness and the production of
at least 400 Perkin Elmer amylase units. The feasibility of increasing the length of time that samples are in the rain simulation
cabinet is being investigated in order to avoid artefacts resulting
from resistance to germination. Advanced lines are also submitted
to the standard dormancy test and the temperature method of dormancy
testing.

Components of sprouting resistance

Of the components contributing to pre-harvest sprouting resistance, namely, dormancy, chaff inhibitors and GA-sensitivity, factors
contributing to dormancy received major attention in the past three
years. Very little is known regarding the relationship between
environment and cultivar as far as sprouting resistance in the South
Western Cape is concerned. However, limited conclusions of these
relationships can be made based on three years' experience at
Bethlehem in the North Eastern Orange Free State. Kernel development normally occurs between mid-October to mid-December. Average
temperatures for this period were calculated from data supplied by
Van der Mey (1980) and cover a range of 31 years. The annual averages vary from 16 to 19°C with an overall mean of 17.5°C.

Temperature ranges and sprouting resistance durations are shown
in Table 1, for locally adapted varieties and breeding lines. Relatively dry conditions preceding harvest ripeness in 1981 may account

for the lower dormancies encountered in this season when compared to the warmer pre-harvest period of 1980. When considered along with the long term temperature values, dormancy levels in local material regularly reach dangerously low levels. Average temperatures were lower than 17°C on seven occasions during the past 31 years. Long term weather data (Van der Mey, 1980) indicates a rainfall of 50 ± 35 mm for the harvesting month, December, further increasing the risk of pre-harvest sprouting.

A number of lines were screened for germination potential in the presence of chaff according to the method of Derera et al. (1977). These lines were grown in the field (Bethlehem) in the relatively cool season of 1979. Six lines exhibited varying levels of germination inhibition in the absence and presence of chaff (Table 2). These lines were tested again in 1980 using seed and chaff from plants grown under hot conditions in a glasshouse. The experimental design was a six by six factorial with three replications (six seed sources and six chaff sources). No significant differences between chaff sources were observed, whereas small but significant differences in response of genotype to the presence of chaff were noted. The possibility that poor agreement of the data obtained in the two years can be ascribed to a temperature effect is being investigated. Using the six lines, crosses in all possible combinations, including reciprocals, were made. Pending development of guidelines for expression of inhibition, the F_1 seed will be used in a genetic study to determine the gene action involved. Near-isogenic lines for this trait will be developed to allow an evaluation of the usefulness of germination inhibition under field conditions.

No data are available concerning GA-sensitivity. However, GA-sensitivity will be studied in the future.

High alpha-amylase levels not due to sprouting

The Springbuck Flats production area (Fig. 1) has summer rainfall. Planting is done in the summer and wheat is harvested during winter months (June to July) when practically no rain falls. Some varieties exhibit undesirably high levels of alpha-amylase despite perfect harvesting conditions.

Conditions existing in the Springbuck Flats were simulated in growth chamber conditions using two varieties exhibiting high levels of alpha-amylase (SST 22 and M12/19) and two normal varieties (Inia and M1/18) in a factorial trial. The day temperature was maintained at 25°C for an eleven hour day. Night temperatures were maintained at either 4°C or 15°C. A second experiment in which exposure to cold conditions (4°C at night and 25°C during the day) at various stages of plant development between fertilization and harvest ripeness was also undertaken.

There was no significant effect associated with temperatures or interaction of temperatures with genotypes. However, genotypes differed significantly at the 1% level. The average alpha-amylase

TABLE 1

Ranges observed in the sprouting resistance of
breeding material in each of 3 years

| Year | Average pre-harvest temperature (°C) | Sprouting resistance (days) | |
		Range observed	Average
1979	16.4	0–30	10.2
1980	17.5	0–15	2.4
1981	17.0	0–15	0.3

TABLE 2

Percentage of seed germinated in the presence of milled bracts

| Line | Germination in presence of bracts[1] (%) | |
	1979	1980[2]
T 78/17	46	91
T 78/4	72	98
T 78/5	75	94
T 77/26	84	97
WSP1/39	100	98
T 78/19	100	93

[1] Expressed as percentage of the viable kernels
[2] Averages from the factorial experiment

values in both of the normal cultivars were 110 and 132 amylase units for Inia and M1/18, respectively. SST 22 and M12/19 had alpha-amylase values of 838 and 1520 units, respectively. In terms of falling number values, these alpha-amylase values correspond to 300-350 seconds for normal cultivars and less than 200 seconds for cultivars with high alpha-amylase.

Preliminary results with the varieties Inia and SST 22 raised the possibility that completely dormant and ungerminated SST 22 may contain a small amount of alpha-amylase isozymes that are normally found in germinated wheat. However, the results were inconclusive and a more comprehensive study of the SST 22 ripening process is planned. The aim is to follow the pattern of isozyme development and degradation in the period between fertilization and harvest ripeness. This will be carried out in a controlled environment using methodology described by Marchylo (1978).

CONCLUSION

Considering the limited research activity in the field of pre-harvest sprouting in South Africa, the main impression would be that local germplasm has a very limited genetic variability for this character. However, from genetic theory, it would seem possible that pure lines might be exceedingly heterogeneous for this character as no selection has been made for sprouting resistance. Directed selection for sprouting resistance in otherwise pure lines might identify individual plants with higher limits of sprouting resistance than would be apparent when considering the bulk population.

The improvement of the pre-harvest sprouting resistance of local germplasm by utilizing the Canadian lines RL 4137 and BW 37 and the Brazilian cultivar Frontana in a back-cross program is currently being implemented.

REFERENCES

Campbell, J.A. 1979. A new method for detection of sprout damaged wheat using a nephelometric determination of alpha-amylase activity. 2nd Int. Sprouting Symp. Cambridge, pp. 107-113.

Derera, N.F., Bhatt, G.M., and McMaster, G.J. 1977. On the problem of pre-harvest sprouting in wheat. Euphytica 26:299-308.

Laubscher, J. 1982. Personal communication. Wheat Board, P.O. Box 908, Pretoria, South Africa.

Marchylo, B.A. 1978. The alpha-amylase isoenzyme system of immature Canadian-grown wheat. Ph.D. thesis, Dept. of Plant Science, University of Manitoba, Canada.

McMaster, G.M., and Derera, N.F. 1975. Methodology and sample preparation when screening for sprouting damage in cereals. Cereal Res. Commun. 4:251-254.

Noll, J.S., and Czarnecki, E. 1979. Methods of extending the testing period for harvest-time dormancy in wheat. 2nd Int. Sprouting Symp., Cambridge, pp. 233-238.

Potgieter, J.H. 1979. Wheat quality and grading. Cereal workshop held at the Wheat Board on 16-17 Oct. 1979, Pretoria, South Africa.

Svensson, G. 1975. Screening methods for sprouting resistance in wheat. Cereal Res. Commun. 4:263-266.

Van der Mey, J.A.M. 1980. Report on the sixth Southern African Regional Wheat Evaluation and Improvement Nursery. Small Grain Centre, Bethlehem, South Africa.

Wheat Board. 1972-1980. Annual reports, P.O. Box 908, Pretoria, South Africa.

Inheritance Studies in Dormancy in Three Wheat Crosses

G. M. Bhatt, F. W. Ellison and D. J. Mares,
*Plant Breeding Institute, The University of
Sydney, Narrabri, New South Wales, Australia*

ABSTRACT

Inheritance of seed dormancy was studied in three crosses of bread wheats (*Triticum aestivum* L.) involving two non-dormant cultivars, Gamut and Shortim and two dormant cultivars Kenya 321 sib and Ford. The parents (P_1 and P_2), F_1's (reciprocals included), F_2's and backcrosses of F_1's to P_1 and P_2 were tested for seed dormancy. Dormancy was found to be controlled by two recessive genes and the segregation ratio of 15 non-dormant : 1 dormant was obtained in the F_2 of two out of three crosses. The two factor segregation ratio was confirmed from the segregation pattern that occurred in the backcross populations. Kenya 321 sib possessed a higher level of dormancy than Ford. Implications of the results are discussed in relation to breeding for pre-harvest sprouting damage resistance.

INTRODUCTION

In the second international sprouting symposium Bhatt and Derera (1979) reported preliminary results on the variability for seed dormancy in wheat afforded by Kenya 321 sib. Also Derera *et al.* (1977) identified the white wheats Kenya 321 sib and Ford as showing a relatively high degree of tolerance to sprouting damage. These findings prompted us to undertake detailed studies on the inheritance of seed dormancy in Kenya 321 sib and Ford. The objective of this paper is to report the results of these studies.

MATERIALS AND METHODS

The experimental populations were derived from the crosses and backcrosses involving four wheat (*Triticum aestivum* L.) cultivars: Kenya 321 sib, Ford, Gamut and

Shortim. The first two cultivars are white wheats with a relatively high degree of tolerance to pre-harvest sprouting damage (Derera *et al.*, 1977). Gamut and Shortim are white wheats of limited tolerance to pre-harvest sprouting damage (Bhatt and Derera, 1979; Derera *et al.*, 1977). The three crosses studied for inheritance of seed dormancy were:

   I   Gamut x Kenya 321 sib
  II   Shortim x Kenya 321 sib
 III   Shortim x Ford

The parents (P_1 and P_2), F_1's (P_1 x P_2), F_1's reciprocal (P_2 x P_1), F_2's (F_1's selfed), B_1 (backcross of P_1 x P_2 to P_1) and B_2 (backcross of P_1 x P_2 to P_2) of each cross were grown in a randomized complete block field experiment at the University of Sydney's Plant Breeding Institute Narrabri during 1981. The three crosses were sown as three experiments. Each experiment consisted of two replications and each replication was composed of 20 experimental rows as follows:

Population	Number of rows
P_1	2
P_2	2
F_1	1
F_1 (reciprocal)	1
F_2	10
B_1	2
B_2	2

Fifteen plants were raised in each row with an intra-row spacing of 10 cm. The rows were spaced 50 cm apart. When the plants attained harvest ripeness, they were harvested individually and the grains were stored at $-10^{\circ}C$ ten days after harvest ripeness until the dormancy tests were undertaken.

Before testing for dormancy, the material was allowed to thaw at room temperature. Fifty seeds per sample were placed between two Whatman No. 1 filter papers in sterile petri dishes and 5 ml of deionized water was added to each petri dish. The petri dishes were then placed in a germination cabinet maintained at $22^{\circ}C$. The germination was allowed to proceed in the dark. The number of seeds that germinated was recorded after a period of 5 days and removed from the petri dishes. The petri dishes containing the ungerminated seeds were then transferred to a germination cabinet maintained at $5^{\circ}C$ in order to break the dormancy of ungerminated seeds. After 4 days in this cabinet, whatever seeds did germinate were counted.

The percentage dormancy of each sample was calculated as the number of seeds which germinated at $5^{\circ}C$ divided by the total number of seeds germinated.

276

RESULTS AND DISCUSSION

In the dormancy tests, plants showing a percentage dormancy of less than 50 were classified as non-dormant, while those showing 50 or more were considered dormant. The results of these tests conducted 10 days after harvest ripeness on the parental as well as hybrid generation material of the three wheat crosses are summarized in Table 1. This criterion of dormancy classified all plants of the parents and F1's into respective dormant or non-dormant groups. Hence the results presented in Table 1 are the combined results from the two replicates.

Parents and F1 Generation.

It is apparent from Table 1 that each of the three crosses involved a dormant and a non-dormant parent. The range in levels of dormancy of the two dormant parents were 50-67% for Kenya 321 sib and 50-56% for Ford, while the two non-dormant parents Gamut and Shortim gave 0% dormancy in our tests.

The F1's of each cross were non-dormant suggesting that the non-dormant condition was dominant to that of dormancy. This was also the case with reciprocal F1's. Lack of reciprocal differences in the three sets of crosses suggests the absence of maternal effects.

Table 1. Classification of wheat plants in parental and hybrid generations of the three wheat crosses as non-dormant types on the basis of dormancy tests conducted 10 days after harvest ripeness.

Generation	Number of Plants					
	Cross I		Cross II		Cross III	
	ND*	D*	ND	D	ND	D
P1	60	0	60	0	60	0
P2	0	60	0	60	0	60
F1	30	0	30	0	30	0
F1**	30	0	30	0	30	0
F2	280	20	283	17	291	9
B1	60	0	60	0	60	0
B2	47	13	44	16	52	8

*ND = non-dormant   *D = Dormant

**F1 = reciprocal F1

Cross I = Gamut (P1) x Kenya 321 sib (P2)
Cross II = Shortim (P1) x Kenya 321 sib (P2)
Cross III = Shortim (P1) x Ford (P2)

F_2 Generation

The paucity of dormant types in the segregating population confirmed the results of F_1's which indicated that dormancy was governed by recessive factors. There was no evidence of transgressive segregation for dormancy in the three crosses. The segregates of two crosses, Gamut x Kenya 321 sib and Shortim x Kenya 321 sib (Table 1) showed a good fit to a duplicate factor ratio of 15 non-dormant : 1 dormant (Table 2), and supports a hypothesis of 2 major gene differences between the parents involved in each cross. Designating the two loci as A and B, a genotype for Gamut and Shortim may be given as AABB and that for Kenya 321 sib as aabb. In the cross, Shortim x Ford, the probability of the segregation occurring in the ratio 15 non-dormant : 1 dormant was low (Table 2). The variation of the observed types from the expected in this cross may indicate the action of modifiers in addition to action of two major genes.

Backcross Generations

On the basis of duplicate recessive factors governing dormancy, the B_1 and B_2 generations should give segregation ratios of 4 non-dormant : 0 dormant and 3 non-dormant : 1 dormant, respectively. The observed B_1 generation data for the 3 crosses given in Table 1 agree with the expectation in this generation since no dormant type was recovered. The observed backcross data for B_2 (Table 1) were tested against the expected by the Chi-square tests and the results are presented in Table 2. The B_2 data for the two crosses, Gamut x Kenya 321 sib and Shortim x Kenya 321 sib showed a good fit to 3 : 1 ratio, which confirmed the F_2 segregation ratio of 15 : 1. In case of Shortim x Ford cross, the segregation pattern showed a poor fit to 3 : 1 ratio, which may be due to the effect of modifiers as hypothesized in the F_2 segregation ratio of this cross.

Table 2. Goodness-of-fit tests for observed vs expected non-dormant : dormant types in the three wheat crosses

Cross	F_2 Generation			B_2 Generation		
	Ratio Tested	Chi-Square	P	Ratio Tested	Chi-Square	P
Cross I	15:1	0.089	0.900-0.750	3:1	0.356	0.500-0.250
Cross II	15:1	0.174	0.750-0.500	3:1	0.089	0.900-0.750
Cross III	15:1	5.408	0.025-0.010	3:1	4.356	0.050-0.025

Cross I = Gamut x Kenya 321 sib. Cross II = Shortim x Kenya 321 sib. Cross III = Shortim x Ford.

278

Implications in Breeding

The two white wheats, Kenya 321 sib and Ford were used as sources of seed dormancy in the present study. Of the two sources, the dormancy of Ford appears marginal (up to 56%) while a moderate level of dormancy was afforded by Kenya 321 sib (up to 67%). The recessive nature of two major gene control of dormancy and the lack of transgressive segregates for dormancy found in this study would imply that a relatively small number of dormant segregates are recoverable from crosses to either of the two sources of dormancy. Large hybrid populations need to be created and intensively selected if dormancy factors of Kenya 321 sib and Ford are to be incorporated into white segregates of acceptable grain quality, high yield and high level of disease resistance.

The limited number of sources of dormancy in white wheats and their nature of inheritance highlights a need to screen white wheats for additional sources of dormancy. A program of this nature has recently been initiated at the Plant Breeding Institute, Narrabri with some 3,000 white seeded cultivars being evaluated in the current year. Clearly, the two most important aspects that could hasten the breeding process aiming at evolving white dormant wheats with adequate dormancy are: high level of dormancy in the parent material and a situation where dormancy is controlled by dominant factors.

ACKNOWLEDGEMENTS

The authors wish to thankfully acknowledge the financial support provided by the Wheat Industry Research Committee of New South Wales.

REFERENCES

Bhatt, G.M., and Derera, N.F. 1979. Potential use of Kenya 321-type dormancy in a wheat breeding program aimed at evolving varieties tolerant to pre-harvest sprouting. Cereal Res. Commun. 8:291-295.
Derera, N.F., Bhatt, G.M., and McMaster, G.J. 1977. On the problem of pre-harvest sprouting of wheat. Euphytica 26:299-308.

The Response of Triticale and Related Cereals to Conditions Inducing Preharvest Sprouting

J. M. McEwan and R. M. Haslemore,
DSIR, Palmerston North, New Zealand

SUMMARY

Grain samples of cultivars of bread wheat, durum, Triticale, and cereal rye exposed to sprouting conditions by delayed harvesting were assessed for sprouting damage by the Hagberg Falling Number Method and by α-amylase estimation. Triticale lines in the trial generally showed high amounts of α-amylase activity throughout the test period. An International Nursery of Triticale breeding stocks was also screened for α-amylase. Some selections with low enzyme content may be useful as parental stocks for this character.

INTRODUCTION

The prevalence in New Zealand of conditions conducive to preharvest sprouting of wheat has previously been reported (McEwan, 1976) and plant breeding measures used to reduce the incidence of sprouting discussed. The use of sequentially harvested trial plots of wheat cultivars and breeding stocks exposed to normal weathering (McEwan, 1980) has enabled a ranking of this material with regard to resistance to preharvest sprouting using the Hagberg Falling Number Method or α-amylase levels as assays of grain damage.

An extension of the cereal breeding programme in the North Island of New Zealand to include durum wheat and Triticale has led to the inclusion of some of these lines in sprouting resistance trials. Triticale characteristically has a higher level of α-amylase than bread wheats at the grain maturity, this being associated with premature cessation of starch deposition and a tendency to produce shrivelled grain (Klassen et al., 1971; Shealy and Simmonds, 1973). These defects have received consideration in plant breeding programmes aimed at the improvement of Triticale (Anon, 1977). High intrinsic levels of α-amylase in the unsprouted grain can cause limitations in its value in cereal processing, while predisposition to sprouting reduces the viability of the grain.

A study of variation in α-amylase levels in
Triticale (Chojnacki et al., 1975) indicated a marked
range in enzyme level with the lowest amount being found
in Triticale derived from the bread wheat cvar. Tom Thumb.
These octoploid Triticale, however would not be of
immediate value in improving the kernel constitution of
the more common hexaploid Triticale.

MATERIALS AND METHODS

Results are presented from field trials carried out
in two seasons when conditions favoured sprouting.
A number of bread wheat lines of known sprouting response
were tested alongside a range of Triticale and representatives of durum wheat and cereal rye as examples of the
progenitors from which hexaploid Triticale has been
derived. All lines were of spring habit with a narrow
range of relative maturity (9 days at ear emergence)
ensuring that the grain samples tested were near the same
stage of physiological maturity at the onset of sprout
inducing conditions. In each season grain samples were
taken at harvest ripeness and subsequently over a period
of up to two months, particularly after rain. Samples
were tested for sprouting damage by the Hagberg Falling
Number Test (AACC Method) for both seasons, though
results are only available for 1979-80 and for α-amylase
levels by the Phadebas α-amylase method (Barnes and
Blakeney, 1974), with the activity expressed as milli
International Enzyme Units per gram of material (m EU/g).
In the initial season grain samples were assessed for
relative test weight, kernel weight and seed germination
by standard laboratory test.

To establish the range of useful variability for
α-amylase level in hexaploid spring Triticale, the 172
entries comprising the 1981 International Triticale
Screening Nursery (ITSN), distributed by CIMMYT, Mexico
were tested, as were a number of breeding stocks
previously selected from CIMMYT material. Test weight
values were also established for the latter lines.

RESULTS

Data from trials in 1979-80 and 1981-82 are
presented. In these seasons periodic rain at harvest
time associated with warm temperatures induced sprouting
in susceptible material. Weather information for these
seasons is summarised in Tables 1 and 3. For 1979-80
bread wheat cultivars were grouped for their responses
to sprouting conditions and the pooled data for α-amylase
activities (Table 1) and Hagberg Falling Number Values
(Table 2) were consistent with previous experience. The
single durum stock tested most closely resembled the

white grained bread wheats in response but the Triticale had high α-amylase levels and low Falling Numbers through-out the test. Associated with the enzymatically induced damage to the grain were overall losses of test weight for all stocks from 74.4 Kg/hl at first harvest to 64.9 Kg/hl at eighth harvest (i.e. 86.5% of original), a reduction of kernel weight of 4% and a decline in germ-ination capacity of the grain from an initial mean of 92%, to 86% for the most resistant cultivar and to 8% for the most susceptible. Rootlet and plumule emergence were apparent at later harvests, particularly in susceptible lines.

TABLE 1. Levels of α-amylase activity in sequentially harvested samples, 1979-80 trial, with weather information

Harvest number	1	2	3	4	5	6	7	8
Cereal group	α-amylase activity (m EU/g)							
Sprouting resistant red grained bread wheats (3 cultivars)	22	20	16	19	19	21	18	73
Moderately sprouting resistant red grained bread wheats (3 cultivars)	21	18	17	17	21	22	20	317
Sprouting susceptible red grained bread wheats (3 cultivars)	36	18	20	17	29	33	30	475
White grained bread wheats (3 cultivars)	28	20	20	23	48	54	80	477
Durum wheat (1 cultivar)	73	35	29	124	240	176	164	416
Triticale (1 cultivar)	260	147	184	248	352	251	304	560
Weather information								
Harvest date (1980)	15-2	22-2	25-2	29-2	4-3	13-3	17-3	9-4
Cumulative Rainfall	-	16.6	23.2	26.3	48.3	57.3	81.2	185.9
No. Raindays (cumulative)	-	2	3	5	9	11	13	26
Mean maximum daily temperature per harvest period (^{o}C)		23.2	20.6	19.7	20.4	21.5	18.4	19.0

TABLE 2. Falling number values for sequentially harvested samples, 1979-80 trial

Harvest number	1	2	3	4	5	6	7	8
Cereal group	Hagberg Falling No. (secs)							
Sprouting resistant red grained bread wheats (3 cultivars)	360	414	424	419	430	384	348	121
Moderately sprouting resistant red grained bread wheats (4 cultivars)	273	333	421	427	435	380	342	81
Sprouting susceptible red grained bread wheats (3 cultivars)	272	421	429	407	262	290	182	60
White grained bread wheats (3 cultivars)	349	379	367	337	254	230	186	65
Durum wheat (1 cultivar)	214	380	416	251	165	176	81	60
Triticale (1 cultivar)	64	71	63	63	63	64	61	60

In the 1981-82 trial, an increased number of Triticale and durum stocks were included, together with cereal rye for the first time. Though the total sampling period was shorter, higher levels of sprouting damage as measured by α-amylase activity were observed (Table 3). Bread wheat and durum lines showed responses similar to those previously observed while the mean value for the 11 Triticale lines was much higher than the bread wheats, as before. Even the Triticale showing lowest levels of activity (Beagle) was considerably above the highest bread wheats. The two stocks of cereal rye resembled the white grained bread wheats in enzyme level, being considerably below the Triticale.

TABLE 3. Levels of α-amylase activity in sequentially harvested samples, 1981-82 trial with weather information

Harvest No.	1	2	3	4	5	6	7	8
	α-amylase activity (m EU/g)							
Sprouting resistant and moderately sprouting resistant red grained bread wheat (2 cultivars)	15	16	18	12	82	426	953	1307
Sprouting susceptible red grained bread wheats (3 cultivars)	16	19	19	17	161	567	1312	2458
White grained bread wheats (2 cultivars)	22	25	24	25	71	444	1693	2033
Durum wheat (2 cultivars)	49	63	181	75	299	221	1241	2397
Triticale (11 cultivars)	786	710	954	1085	1473	1852	3037	4848
Cereal rye (2 cultivars)	27	39	16	91	21	349	2322	3293
Weather information								
Harvest date (1982)	1-2	5-2	8-2	12-2	22-2	1-3	5-3	12-3
Cumulative Rainfall (mm)	-	0.4	8.0	8.0	24.5	52.7	100.6	100.6
No. Raindays (cumulative)	-	1	2	2	5	9	12	12
Mean maximum daily temperature per harvest period (°C)		23.4	20.5	25.5	22.5	24.7	20.2	23.1

Samples from the 172 lines of the 1981 ITSN were
tested for α-amylase level using Beagle as a local
Triticale standard (Fig. 1). Twenty-two selections
recorded enzyme levels below 600 m EU/g as compared with
the mean score for Beagle of 780 units. None of these
lines, however, reached the low levels of the bread
wheat Karamu, with a mean value of 24 units from
5 samples. Some of the lines with very high levels of
activity showed signs of incipient sprouting. A group of
25 Triticale breeding stocks of CIMMYT origin undamaged
by sprouting gave no correlation between α-amylase level
and test weight (r = -.091).

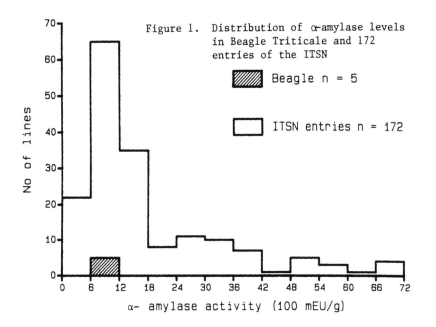

Figure 1. Distribution of α-amylase levels
in Beagle Triticale and 172
entries of the ITSN

Beagle n = 5

ITSN entries n = 172

DISCUSSION

The trial system adopted by Crop Research Division,
Palmerston North to assess relative resistance to pre-
harvest sprouting has given consistent results over
a number of seasons when conditions that induced sprouting have occurred. This information is of value in the
cereal breeding programme as even a short period of seed
dormancy may be effective in protecting commercial crops.
The results in the trials discussed here are in accord
with previous observations in that the red grained bread
wheats had a range of sprouting resistance while the
white grained lines were susceptible. The Triticale
lines had high levels of α-amylase activity compared with
the representatives of durum wheat and cereal rye tested

as examples of the stock from which hexaploid Triticale was derived. The anomalous endosperm development in Triticale described by Klassen et al., (1971) and Shealy and Simmonds (1975), revealed a premature termination of starch deposition leading to shrivelled kernels of low test weight. A substantial increase in α-amylase levels in maturing kernels of some Triticale lines was also reported, leading to a predisposition to sprouting. High amylase levels are also undesirable in some commercial processes for which Triticale grain could be considered such as starch manufacture.

Plant breeding projects aimed at improving Triticale have included selection for improved grain filling capacity. Klassen et al., (1971) reported an inverse correlation between α-amylase level and grain density but a similar relationship between α-amylase level and test weight was not found in this study. The samples tested from the 1981 ITSN displayed a wide range of α-amylase levels. Lines with particularly high enzyme content frequently showed visible sprouting damage. None of the lines reached the low levels determined for Karamu bread wheat, but a number were below the level of Beagle Triticale. Of the 12 stocks tested with lowest activity 6 included the CIMMYT selection Panther "S" in the parentage. This stock may provide a useful genetic source for low α-amylase level parental stock in Triticale. Continued breeding and selection for superior grain filling in Triticale could result in a correlated response of reduced α-amylase levels should the early cessation of starch deposition in the endosperm be associated with the increase in α-amylase level.

ACKNOWLEDGEMENTS

The authors are grateful to Mr L.D. Simmons of Wheat Research Institute, DSIR, Christchurch for permission to use the Falling Number Values in Table 2.

REFERENCES

Anon. 1977. CIMMYT Report on Wheat Improvement 1977. El Batan Mexico, 245 pp.
Barnes, W.C. and Blakeney, A.B. 1974. Determination of cereal alpha amylase using a commercially available dye-labelled substrate. Die Stärke 26: 193-197.
Chojnacki, G., Brykczyinski, J. and Tymieniecka, E. 1976. Preliminary information on sprouting in Triticale. Cereal Res. Comm. 4: 111-114.
Klassen, A.J., Hill, R.D. and Larter, E.N. 1971. Alpha-amylase activity and carbohydrate content as related to kernel development in Triticale. Crop Sci. 11: 265-267.
McEwan, J.M. 1976. Breeding for resistance to pre-harvest sprouting in New Zealand wheats. Cereal

Res. Comm. 4: 97-100.

McEwan, J.M. 1980. The sprouting reaction of stocks with single genes for red grain colour derived from Hilgendorf 61 wheat. Cereal Res. Comm. 8: 261-264.

Shealy, H.E. and Simmonds, D.H. 1973. The early developmental morphology of Triticale grain. Proc. 4th Int. Wheat Genetics Symp. Columbia, Missouri, 265-270.

Wheat Weathering Damage in Windrower/Combine vs. Direct-Combine Harvesting Systems

J. M. Clarke, *Research Station, Agriculture Canada, Swift Current, Saskatchewan, Canada*

SUMMARY

Weathering damage was studied in standing and wind-rowed wheat in both plot-scale and field-scale trials. The plot trials were conducted with a range of dormant and nondormant red- and white-grained wheats. Falling number was markedly lower in windrowed than in standing material, particularly in nondormant genotypes. These differences were readily apparent with total rainfall as low as 49 mm during the weathering period. Similar results were recorded in field-scale trials. The difference in weathering susceptibility between standing and windrowed material was attributed to higher moisture contents and slower drying following rains in windrowed compared to standing crops.

INTRODUCTION

In the prairie region of Canada cereal crops are normally cut with a windrower when the grain is in the 30 to 40% moisture content range. The windrows are then picked up and threshed by combine harvesters when the grain has dried to safe storage moisture levels.

The practice of windrowing cereal crops prior to combine harvesting arose for several reasons. Briefly, these can be stated as: (1) hastening the final drying of the crop to permit earlier threshing, which would reduce exposure to damaging weather; (2) in the case of wheat, to reduce losses caused by the wheat stem sawfly (Cephus cinctus Nort.) in areas where it is a common problem; (3) to avoid storage problems caused by immature weed seeds in the grain; and (4) to reduce shattering due to natural and mechanical causes (Dodds 1967). Modern herbicides have largely eliminated the green weed problem, while wheat stem sawfly losses can be very much reduced by growing solid-stemmed wheat cultivars.

The continued use of windrowing then hinges on whether, in fact, windrowing hastens drying and reduces shattering losses. Research is underway in both of these areas. Weathering damage to windrowed crops compared to other methods of harvest was studied somewhat during the 1930s, but has been largely ignored until recently. Harrington (1932) found that sprouting tended to be greater in windrowed than in stooked wheat. Prolonged wet periods with little or no drying were particularly damaging. More recently, Briggle (1980) reported that early harvest coupled with the use of grain dryers has reduced sprouting damage in Michigan wheat crops.

Few references to weathering damage in windrowed wheat can be found. The objective of the work reported here was to compare weathering damage in standing and windrowed red- and white-grained sprouting resistant and susceptible genotypes of wheat.

MATERIALS AND METHODS

Studies of Triticum aestivum L. genotypes were conducted in small research plots using simulated windrows as well as in larger field plots harvested by prototype windrowers. Data are reported for small-plot studies conducted at Beaverlodge, Alberta in 1978 and 1981, and at Swift Current and Regina, Saskatchewan in 1981. Field-scale data are reported for Swift Current in 1980 and 1981.

The 1978 trial at Beaverlodge was arranged as a split-plot in randomized complete block with four replications. The main plots were a factorial of two genotypes, Neepawa (red-grained) and NB 112 (white-grained), and two cutting treatments, standing vs. windrowed. Subplots consisted of six sampling times. At each sampling, commencing at approximately 35% kernel moisture, designated plots were cut and laid on the stubble (windrowed treatment) or cut, threshed and artificially dried (standing treatment). The windrowed material was collected and threshed when dry. Falling number and test weight were determined in the threshed grain.

In 1981, the small-plot trial at Beaverlodge, Swift Current, and Regina was a split-plot in randomized complete block design with four replications. Main plots consisted of fourteen genotypes (eight red, six white) of varying postharvest dormancy, while the subplots consisted of a nonweathered (early) sample and standing and windrowed weathered samples (late). The nonweathered samples were collected at about 25 to 35% kernel moisture, at which time the windrowed treatment was cut and laid on the stubble. The weathered samples were collected eight to ten weeks later. Falling number was determined on all samples.

The 1980 field-scale trial was a split-plot in randomized complete block design with five replicates utilizing the genotype Neepawa. Main plots were harvest
methods, windrowed vs. standing, while subplots were
time of sampling. The windrowed treatment was cut with
a 6.1-m wide windrower when the crop was about 35% kernel moisture content. Samples of both the windrowed and
standing material were taken over a period of eight
weeks. Falling number, test weight, and commerical
grades were determined.

The 1981 field-scale trial was similar, except that
the genotypes Columbus, Kenya 321, and NB 106 were included.

RESULTS AND DISCUSSION

There was a marked difference in falling number of
windrowed and standing wheat at Beaverlodge in 1978
(Table 1). Exposure of windrows to 136 mm of rain between 11 August and 26 September reduced falling number
to 62 in both NB 112 and Neepawa. The reduction in the
falling number of the standing treatment, 11 August compared to 26 September, was much less. Falling number of
Neepawa was affected less than that of NB 112.

Table 1
Effect of weathering on falling number of windrowed and
standing NB 112 and Neepawa wheat in plot-scale tests at
Beaverlodge, Alberta in 1978

Cutting date[+]	Precip. between cuts (mm)	NB 112		Neepawa	
		Wind-rowed	Stand-ing	Wind-rowed	Stand-ing
11 Aug.		62	425	62	560
	56				
22 Aug.		62	492	162	545
	15				
31 Aug.		62	303	418	511
	65				
26 Sept.		254	231	497	443

$LSD_{.05}$ = 76 for harvest method means
[+]Date cut and laid on stubble (windrowed treatment) or
cut, threshed and dried (standing treatment)

There was no genotypic difference in test weight
loss, but test weight loss was much greater in windrowed
than in standing material (data not shown). In the
longest-exposed material, test weight was 65 kg/hL in
windrows compared to 75 kg/hL in the standing crop. Low

Table 2
Falling numbers of weathered and nonweathered standing and windrowed dormant and nondormant red and white wheats in plot-scale trials in 1981 (means of three locations)

Class	Genotype	Falling number (sec)			
		Non-weathered	Weathered Standing	Weathered Windrowed	Mean
Red Dormant	RL 4137	515	433	277	408 ab
	Columbus	586	472	275	445 a
	Leader	464	540	227	411 ab
	Neepawa	534	471	275	400 abc
Nondormant	Park	497	439	147	361 bcde
	Cypress	468	507	125	367 bcd
	Chester	447	373	135	318 def
	Garnet	400	358	70	276 f
White Dormant	Kenya 321	410	475	254	380 bc
	CT 932	440	423	235	366 bcd
	NB 402	450	403	174	342 cde
Nondormant	Fielder	372	325	133	277 f
	NB 106	458	380	109	316 def
	NB 112	476	355	76	302 ef
	Mean	465 a	425 b	173 c	

Genotype or harvest method means followed by the same letter do not differ significantly according to Duncan's Multiple Range Test (P = .05)

test weight would have placed the windrowed material in the lowest commercial grade (Anonymous 1978).

In a broader range of genotypes tested at three locations in 1981, there was again a marked difference between falling numbers of standing and windrowed material (Table 2). The differences between the standing and windrowed treatments were more pronounced within the nondormant than within the dormant groupings of genotypes.

Orthogonal single degree of freedom comparisons within the windrowed treatment indicated no significant difference between the red and white genotypes. Within color classes, however, the differences between the dormant and nondormant groupings were significant. These results contrast with earlier studies which indicated that white-grained wheats are more sprouting susceptible than red-grained wheats (e.g., Wellington and Durham 1958). There is clearly considerable overlap of dormancy of the color classes.

Precipitation differences at the three locations affected treatment differences in falling number (Table 3). Higher precipitation at Regina than at the other locations resulted in significantly lower falling numbers in both late-sampled standing and windrowed material compared to the nonweathered material. As little as 49 mm of rain provided a good differential between standing and windrowed material at Swift Current.

Table 3
Precipitation and falling number of nonweathered and weathered standing and windrowed wheat (mean of 14 genotypes) at three locations in 1981 (plot-scale trials)

Location	Precip. (mm)	Non-weathered	Weathered	
			Standing	Windrowed
Beaverlodge	78	378 a	364 a	200 b
Regina	105	325 a	250 b	115 c
Swift Current	49	693 a	661 a	206 b

Means within locations followed by the same letter do not differ significantly according to Duncan's Multiple Range Test (P = .05)

Field-scale trials with Neepawa showed greater reduction in test weights of windrowed than of standing material (Table 4). Similarly, falling number was

Table 4
Effect of weathering on test weight and falling number
of standing and windrowed Neepawa wheat (1980 field-
scale plots)

Date of sampling	Test weight (kg/hL)		Falling number (sec)	
	Windrowed	Standing	Windrowed	Standing
29 Aug.	80 ± 0.3[+]	80 ± 0.3	399 ± 13	385 ± 17
6 Sept.	80 ± 0.1	80 ± 0.3	398 ± 10	394 ± 14
7 Oct.	76 ± 0.1	76 ± 0.5	307 ± 7	323 ± 6
17 Oct.	69 ± 0.9	73 ± 0.5	85 ± 14	247 ± 5

[+]Standard error of the mean

affected by weathering to a greater degree in the windrowed than in the standing treatment. Commercial grade
was one level higher in standing than in windrowed material (#2 vs. #3). Falling number differences between
windrowed and standing material were greater in the non-
dormant genotype NB 106 than in Columbus, Neepawa and
Kenya 321 in 1981 (Table 5).

Table 5
Falling number of weathered and nonweathered standing
and windrowed wheat (field-scale plots 1981)

Genotype	Falling number (sec)		
	Nonweathered	Weathered	
		Windrowed	Standing
Columbus	439 ± 16[+]	330 ± 30	386 ± 51
Neepawa	477 ± 21	311 ± 12	392 ± 28
Kenya 321	434 ± 21	331 ± 17	334 ± 48
NB 106	460 ± 7	118 ± 15	304 ± 33

[+]Standard error of the mean

Much of the difference in weathering damage between
windrowed and standing wheat probably arises from the
difference in drying rates following rain. In 1980, for
example, moisture contents of both grain and straw were
much higher in the windrowed than in the standing material following a light rain (Table 6). Higher grain
moistures, maintained for longer periods in the windrow
than in the standing crop, provide more time for alpha-
amylase generation.

Table 6
Grain and straw moistures of standing and windrowed
Neepawa wheat following rain (1980 field-scale plots)

Component	% Moisture	
	Windrowed	Standing
Grain	$34.0 \pm 2.5^+$	20.7 ± 0.5
Straw	51.2 ± 1.0	29.5 ± 2.1

+Standard error of the mean

There is clearly potential for greater weather
damage in windrowed than in standing wheat, even with
relatively low rainfall amounts. Screening of genotypes
for weathering resistance in the field should, therefore, be done in simulated windrows in order to get a
good differential between genotypes. There is a need
for further study of weathering damage in field-scale
windrows, particularly under higher rainfall conditions,
which would allow a better assessment of commercial
grade differences.

REFERENCES

Anonymous. 1978. Specifications for official grades of
Canadian Grain. Canadian Grain Commission, Winnipeg, Canada.
Briggle, L.W. 1980. Pre-harvest sprout damage in wheat
in the U.S. Cereal Res. Comm. 8: 245-250.
Dodds, M.E. 1967. Wheat harvesting - machines and
methods. In Proc. Canadian Centennial Wheat Symposium, K.F. Nielsen (Ed.), Modern Press, Saskatoon.
Harrington, J.B. 1932. The comparative resistance of
wheat varieties to sprouting in the stook and windrow. Sci. Agri. 12: 635-645.
Wellington, P.S. and Durham, V.M. 1958. Varietal
differences in the tendency of wheat to sprout in
the ear. Emp. J. Exp. Agric. 26: 47-54.

Participants in the Third International Symposium on Pre-Harvest Sprouting in Cereals

1. Paul Kereru, 2. Ian Gordon, 3. Peter Meredith, 4. Ann Oaks, 5. Margret Ogolla, 6. Maria Ortega-Delgado, 7. Jim Kruger,
8. Mike Gale, 9. Solomon Kibite, 10. Alejandro Blanco, 11. Wilma Kruis, 12. Nick Derera, 13. George Freeman, 14. Volkmar Stoy,
15. Rod King, 16. Erling Strand, 17. Carol Duffus, 18. Franklin Fong, 19. John Noll, 20. Brian Marchylo, 21. Jorge Molina,
22. Ferdinand Weilenmann, 23. Andy Cairns, 24. Mary Hagemann, 25. Lars Reitan, 26. Peter Nicholls, 27. Jennifer Mitchell,
28. Joel Dick, 29. Bryan Harvey, 30. Richard Frohberg, 31. Estela Sanchez de Jimenez, 32. Gerry Thraves, 33. Allan Ciha,
34. Gregory Gibbons, 35. Drusilla Pearson, 36. Edward Liebe, 37. Don LaBerge, 38. Stig Larsson, 39. Kare Ringlund,
40. Sandy MacGregor, 41. Randy Weselake, 42. Fred Townley-Smith, 43. Bill Woodbury, 44. Bob Hill, 45. Roy Cantrell,
46. Martin McEwan, 47. Felix Mederick, 48. Don Salmon, 49. Brian Rossnagel, 50. Grant McLeod, 51. Tom McCaig, 52. Ron DePauw,
53. John Clarke, 54. Judith Fregeau, 55. Edward Czarnecki, 56. Daryl Mares.

Missing: Elaine Asp, Hans Lindblom, Peter Ranum.

Participants

ASP, ELAINE, Dept. of Food Science and Nutrition, University of Minnesota, 1334 Eckles Ave., St. Paul, MN 55208, U.S.A.

BLANCO, ALEJANDRO, Fac. Química (D.E.S.), Universidad Nal. Autónoma de México, México 20, D.F., México

CAIRNS, A.L.P., Dept. of Agronomy, University of Stellenbosch, Stellenbosch, South Africa

CANTRELL, Roy G., North Dakota State University, Agronomy Dept., Fargo, ND 58105, U.S.A.

CIHA, ALLAN J., Dept. of Agronomy, USDA-ARS, 209 Johnson Hall, Washington State University, Pullman, WA 99164, U.S.A.

CLARK, PHILLIP, Agriculture Canada, Beaverlodge Research Station, Box 29, Beaverlodge, Alberta, Canada T0H 0C0

CLARKE, JOHN M., Agriculture Canada, Box 1030, Swift Current, Saskatchewan, Canada S9H 3X2

CZARNECKI, EDWARD, Agriculture Canada, 195 Dafoe Rd., Winnipeg, Manitoba, Canada R3T 2M9

DE PAUW, RON, Agriculture Canada, Box 1030, Swift Current, Saskatchewan, Canada S9H 3X2

DERERA, NICK, Agricultural Science Consultant, 5 Lister St., Winston Hills 2153, N.S.W., Australia

DICK, JOEL W., Dept. of Cereal Chemistry and Technology, North Dakota State University, Harris Hall, Fargo, ND 58105, U.S.A.

DUFFUS, CAROL M., School of Agriculture, University of Edinburgh, West Mains Rd., Edinburgh, EH9 3JG, Scotland

FONG, FRANKLIN, Plant Sciences Dept., Texas A&M University, College Station, TX 77843, U.S.A.

FREEMAN, W. GEORGE, President, Prime Wheat Association, P.O. Box 146, Narrabri, N.S.W., Australia

FRÉGEAU, JUDITH, Agriculture Canada, Building 75, Ottawa, Ontario, Canada, K1A 0C6

FROHBERG, RICHARD C., Agronomy Dept., North Dakota State University, Fargo, ND, 58105, U.S.A.

GALE, MIKE D., Plant Breeding Institute, Maris Lane, Trumpington, Cambridge, England CB2 1LQ

GIBBONS, GREGORY C., Carlsberg Research Lab., Dept. of Biotechnology, Gamle Carlsberg Vej 10, DK-2500 Copenhagen Valby, Denmark

GORDON, IAN L., Agronomy Dept., Massey University, Palmerston North, New Zealand

HAGEMANN, M.G., Johnson Hall #207, Washington State University, Pullman, WA, 99164, U.S.A.

HARVEY, BRYAN L., Crop Science Dept., University of Saskatchewan, Saskatoon, Saskatchewan, Canada S7N 0W0

HILL, BOB, Plant Science Dept., University of Manitoba, Winnipeg, Manitoba, Canada R3T 2N2

KERERU, PAUL, National Plant Breeding Station, Njoro, Kenya

KIBITE, SOLOMON, University of Alberta, Edmonton, Alberta, Canada T6G 2E1

KING, ROD, CSIRO, Div. of Plant Industry, P.O. Box 1600, Canberra, ACT 2601, Australia

KRUGER, JIM E., Canadian Grain Commission, Grain Research Lab., 1404-303 Main St., Winnipeg, Manitoba, Canada R3C 3G8

KRUIS, WILMA, Small Grain Centre, Private Bag X29, Bethlehem, South Africa

LABERGE, DON, Canadian Grain Commission, Grain REsearch Lab., 1404-303 Main St., Winnipeg, Manitoba, Canada R3C 3G8

LARSSON, STIG, The Swedish Seed Association, S-268 00 Svalöv, Sweden

LIEBE, EDWARD, Foreign Agricultural Service, USDA, Rm. 4932, South Agricultural Bldg., 1400 Independence Ave., S.W., Washington, D.C. 20250, U.S.A.

LINDBLOM, HANS, Dept. of Plant Husbandry, Swedish University of Agricultural Sciences, S-75-007 Uppsala, Sweden

MacGREGOR, A.W. (Sandy), Canadian Grain Commission, Grain Research Lab., 1404-303 Main St., Winnipeg, Manitoba, Canada R3C 3G8

MARCHYLO, BRIAN A., Canadian Grain Commission, Grain Research Lab., 1404-303 Main St., Winnipeg, Manitoba, Canada R3C 3G8

MARES, DARYL J., Plant Breeding Institute, University of Sydney, P.O. Box 219, Narrabri, N.S.W. 2390, Australia

McEWAN, J. MARTIN, Crop Research Division, DSIR, Private Bag, Palmerston North, New Zealand

McCAIG, TOM N., Agriculture Canada, Box 1030, Swift Current, Saskatchewan, Canada S9H 3X2

McLEOD, GRANT J., Agriculture Canada, Box 1030, Swift Current, Saskatchewan, Canada S9H 3X2

MEDERICK, FELIX, Alberta Agriculture, Bag Service #47, Lacombe, Alberta, Canada

MEREDITH, PETER, Wheat Research Institute, P.O. Box 1489, Christchurch, New Zealand

MITCHELL, JENNIFER, Dept. of Plant Science, University of Manitoba, Winnipeg, Manitoba, Canada R3T 2N2

MOLINA, Jorge, Fac. Química (D.E.S.), Universidad Nal. Autónoma de México, México 20, D.F., México

NICHOLLS, PETER B., Waite Agricultural Research Institute, Private Mail Bag #1, Glen Osmond, S.A. 5064, Australia

NOLL, JOHN S., Agriculture Canada, 195 Dafoe Rd., Winnipeg, Manitoba, Canada R3T 2M9

OAKS, ANN, McMaster University, 1280 Main Street West, Hamilton, Ontario, Canada L8S 4K1

OGOLLA, MARGRET, Plant Science Dept., University of Manitoba, Winnipeg, Manitoba, Canada R3T 2N2

ORTEGA-DELGADO, María Luisa, Colegio de Postgraduados, Chapingo, Méx., México

PEARSON, DRUSILLA, Dept. of Plant Science, University of Alberta, Edmonton, Alberta, Canada

RANUM, PETER M., Pennwalt Corporation, 3000 Ranchview Lane, Plymouth, MN 55447, U.S.A.

REITAN, LARS, Voll Agricultural Research Station, Box 1918, Moholtan, N-7001 Trondheim, Norway

RINGLUND, KÅRE, Agricultural University of Norway, Boks 41, 1432 AAS-NLH, Norway

BRIAN ROSSNAGEL, Crop Science Dept., University of Saskatchewan, Saskatoon, Saskatchewan, Canada S7N 0W0

STRAND, ERLING, Dept. of Farm Crops, Agricultural University of Norway, Postbox 41, N-1432 AS NLH, Norway

SALMON, DON F., Alberta Agriculture, Bag Service #47, Lacombe, Alberta, Canada

SÁNCHEZ DE JIMÉNEZ, ESTELA, Dept. de Bioquímica, Fac. de Química, Div. de Estudios Superiores, Universidad Nacional Autónoma de México, México 20, D.F., México

STOY, VOLKMAR, The Swedish Seed Association, S-268 00 Svalöv, Sweden

THRAVES, GERRY, Plant Science Department, University of Manitoba, Winnipeg, Manitoba, Canada R3T 2N2

TOWNLEY-SMITH, T. FRED, Agriculture Canada, Box 1030, Swift Current, Saskatchewan, Canada S9H 3X2

WEILENMANN, FERDINAND, Swiss Federal Research Station for Agronomy, CH-8046 Zurich-Reckenholz, Switzerland

WESELAKE, RANDY, Plant Science Dept. University of Manitoba, Winnipeg, Manitoba, Canada R3T 2N2

WOODBURY, BILL, Dept. of Plant Science, University of Manitoba, Winnipeg, Manitoba, Canada R3T 2N2

Notice of Next Meeting

The Fourth International Symposium on Pre-Harvest Sprouting Damage in Cereals has been tentatively scheduled for January 1986 in Australia. The Committee for that meeting will be: Dr. J. E. Kruger (Canada), President; Dr. D. J. Mares (Australia), Secretary; Mr. N. Derera (Australia), Dr. V. Stoy (Sweden), Dr. F. Weilenmann (Switzerland), and Dr. M. Gale (England). Information concerning the meeting can be obtained from any of the above organizers.

Index

302

312

Printed and bound by CPI Group (UK) Ltd, Croydon, CR0 4YY

23/10/2024

01778241-0018